CHARACTERISTICS OF SEA REVERBERATION

STATISTICHESKIE SVOISTVA MORSKOI REVERBERATSII

СТАТИСТИЧЕСКИЕ СВОЙСТВА МОРСКОЙ РЕВЕРБЕРАЦИИ

Characteristics of
Sea Reverberation

V. V. Ol'shevskii

Acoustics Institute
Academy of Sciences of the USSR, Moscow

Translated from Russian

With a Foreword by
Vernon M. Albers
Ordnance Research Laboratory
Institute for Science and Engineering
The Pennsylvania State University, University Park

 SPRINGER SCIENCE+BUSINESS MEDIA, LLC 1967

V. V. Ol'shevskii was born in 1932 and was graduated from the Radio Communications and Broadcasting Department of the Moscow Communications Electrical Engineering Institute in 1955. In 1965 he presented his thesis and was awarded the degree of Candidate of Engineering Sciences. Currently he is at the Acoustics Institute of the Academy of Sciences of the USSR. His published works relate to research in wave scattering, the statistical analysis of signals, and methods of information processing.

The original Russian text was published for the Acoustics Institute of the Academy of Sciences of the USSR by Nauka Press, Moscow, in 1966.

Виктор Владимирович Ольшевский

СТАТИСТИЧЕСКИЕ СВОЙСТВА
МОРСКОЙ РЕВЕРБЕРАЦИИ

ISBN 978-1-4899-4724-6   ISBN 978-1-4899-4722-2 (eBook)
DOI 10.1007/978-1-4899-4722-2

Library of Congress Catalog Card Number 67-25401

© 1967 Springer Science+Business Media New York
Originally published by Plenum Publishing Corporation, New York in 1967
Softcover reprint of the hardcover 1st edition 1967

FOREWORD

The earliest studies of reverberation in the ocean were made because of its importance as a signal which interfered with the target echoes received by SONAR systems. The early studies demonstrated that, in addition to reverberation due to scatterers at the boundaries, there is reverberation due to scattering from inhomogeneities in the volume of the medium.

Reverberation in rooms is determined by the nature of the signal and the scattering from the boundaries of the room. The scattering elements in rooms can always be directly observed, while those in the ocean, particularly those in the volume of the medium, cannot be directly observed. In addition, when reverberation measurements are made in the ocean, the source and receiver as well as the inhomogeneities in the volume of the medium are often moving. The reverberation in the ocean, therefore, serves as a means of studying the distribution and the nature of scatterers in the ocean and is used by oceanographers as a means of studying the ocean medium and its boundaries. For example, most of our knowledge of the deep scattering layers is derived from measurements of sound scattering from them.

Since World War II, there have been a large number of publications on underwater reverberation measurements and theory. However, there has been no previous systematic presentation of the results of experimental and theoretical investigations on reverberation in the oceans.

In this book, Ol'shevskii has summarized the literature on reverberation measurements and theory and has presented an analysis of the statistical characteristics of sea reverberation on the basis of a discrete model of the sound scattering by inhomogeneities in the medium and irregularities at its boundaries. His presentation also takes into account the nature of the transmitted signal, the bandpass characteristics of the receiver, and motions of the source and the inhomogeneities.

This book should be a valuable aid to those engaged in the development of SONAR systems and it should be of particular value to the oceanographer who uses the results of reverberation measurements to derive information about the scatterers in the volume of the medium and at its boundaries.

Vernon M. Albers

State College, Pennsylvania
May 25, 1967

PREFACE TO THE AMERICAN EDITION

This work presents an analysis of the statistical characteristics of sea reverberation on the basis of a discrete model of sound scattering by inhomogeneities in the ocean medium and irregularities at its boundaries. Despite the inadequacy of such a model in several cases for the description of real phenomena (for example, in the investigation of scattering at the ocean surface or by thermal irregularities), it is expected that the results of the present statistical analysis of reverberation signals will have a sufficiently general character.

The American edition of the monograph closely parallels the edition published in the Soviet Union in 1966.

The author will gratefully acknowledge those readers who would be so kind as to inform him of their comments regarding the book, and their appraisal of it.

V. V. Ol'shevskii

Moscow, December 1966

PREFACE

One of the interesting physical phenomena in underwater acoustics is sea reverberation, the origin of which is related to the scattering of sound by inhomogeneities of the ocean medium and the irregularities of its boundaries. The investigation of the statistical properties of sea reverberation is a vital problem in underwater acoustics today.

In the last ten years a great many papers have been published on the investigation of sea reverberation, including the investigation of its statistical aspects, but this material is not systematic in character and in general lacks a unified approach to the study of sea reverberation as a stochastic process.

The present monograph is devoted to a systematic presentation of the results of theoretical and experimental investigations of the statistical properties of sea reverberation. The need for such a book has long been felt and its publication at this time is dictated both by the requirements of applied underwater acoustics and by the desirability of collecting the end results of more than a decade of research on this complex statistical effect.

The statistical properties of reverberation are studied in this book by means of the methods of statistical communication theory, in particular the theorem of superposition of stochastic perturbations. This mode of presentation was motivated first and foremost by the peculiarities of reverberation as a stochastic process and its analogy with a great many other physical phenomena bearing on certain types of fluctuation processes in electronic engineering.

The indicated approach was determined largely by the profile of specialists for whom the monograph was intended. These include scientists and engineers familiar with the fundamentals of the theory of stochastic functions and specializing in the acoustics of the ocean and methods of investigating the properties of stochastic processes. The monograph should also prove useful to students and degree candidates for their study of underwater acoustical phenomena.

It would be difficult to name all those who have helped me in one way or another to conduct the investigations and prepare the book, and to whom I ought to express my most sincere appreciation. It would be impossible, however, not to thank my colleagues from the Acoustics Institute of the Academy of Sciences of the USSR, M.G. Borisov, G.A. Gorlov, I.I. Dobretsov, E.V. Kirillov, V.E. Kitaiskii, I.P. Konovalov, Yu.V. Lunichkin, G.S. Ol'shevskaya, V.I. Kharchevnikov, V.D. Tsyganko, and L.M. Chibisov, who have at various times since 1955 rendered aid in planning and conducting the sea measurements, processing the experimental data, and organizing the work for publication, as well as V.A. Antonov, L.F. Bondar', V.I. Vorob'ev, V.M. Matangin, S.V. Leshukov, and V.A. Khvorostov for their contribution by way of discussions of the results, calculations, and verification of the equations. I am particularly grateful to V.A. Zakharov, whose consultations and assistance have exerted a profound influence on the manner in which the content of the book is presented, and whose collaboration has not only proved highly productive, but has also brought great satisfaction.

I am also pleased to express my appreciation to Professor V.S. Pugachev for his attention to the book.

I am deeply indebted to my teachers, Professor Yu. M. Sukharevskii, who supervised my research for a number of years, and Professor G. D. Malyuzhinets, whose counsel has always deepened my insight into that which I set down on paper.

The author will gratefully accept any comments regarding the content of the book.

V.V.O.

CONTENTS

CHAPTER I

DESCRIPTION OF THE PROBLEM AND
METHODS OF SOLUTION

§ 1. Introduction

   Probabilistic methods for the analysis of diverse effects, underwater acoustical phenomena included, are important for at least two reasons.

   First, these methods make it possible to investigate the fine structure of processes and to disclose the physical characteristics of effects; they enable one to describe the average and stable laws involved, and to establish (with suitable experimental data available) the degree of correspondence between the adopted model and the actual causes behind the effect.

   Second, statistical treatment often proves the most effective in various applications; for example, the investigation of information processing methods and the analysis of stochastic processes, in particular, as they affect various technical devices require, above all, knowledge of the statistical characteristics of the input signals, a fact which makes the theory of stochastic functions particularly well suited to the investigation of the properties of sea reverberation.

   The analysis of reverberation signals is based on a definite model of the phenomenon, from which it is then possible to calculate the one-dimensional and, in some cases, the multi-dimensional probability distributions of reverberations, its envelope, and phase. The experimental confirmation of at least some of these characteristics either corroborates the adopted model or suggests some refinement of it. Accordingly, one may study the nature of scattering in the ocean, explain the physical properties of the scatterers, and investigate the statistical methods for processing underwater acoustical information.

   The first investigations of reverberation as a stochastic process were carried out by Yu. M. Sukharevskii. In one of his papers [23], published in 1947, inferences were made with regard to the nature of the fluctuations of the reverberation envelope and certain features of sound scattering were indicated, permitting the reverberation process to be regarded to a certain approximation as the sum of a large number of elementary scattered signals.

   Beginning in 1955, the Acoustics Institute of the Academy of Sciences of the USSR undertook some theoretical and experimental investigations of the statistical properties of reverberation with a view toward analyzing its probability distributions, correlation characteristics, and energy spectra for different types of modulation of the transmitted signals. Some of the results of these investigations, carried out mainly up to 1960, are found in [13-18].

   At this time there appeared a number of papers published by foreign authors on the investigation of reverberation in approximately the same aspects. In [25], for example, data are presented on the distribution of the reverberation envelope and its correlation characteristics for the transmission of rectangular pulses with a sinusoidal carrier, along with the results of an analysis of the frequency composition of reverberation signals. Information regarding the energy spectra of reverberation is contained in [37]. Other articles [30, 43, and, in part 41, 42] are devoted to the study of reverberation in a continuous transmission mode.

It is important to mention [27, 37, 41, 42, 53, 55], in which a statistical treatment of underwater echo location (sonar) is used as a basis for the analysis of certain properties of reverberation during the transmission of signals with different types of amplitude and frequency modulation, and general techniques are outlined for minimizing the influence of reverberation noise in sonar. An interesting article in this respect is [53], in which the properties of reverberation are analyzed by means of ambiguity diagrams, the theory of which has been developed in articles on radar (see, e.g., [3]).

Later works [31, 35] deal with the investigation of the statistical characteristics of reverberation, based on the conception of this process as a sum of stochastic perturbations.

It is quite natural that the results to date should allow a generalization of the cumulative experience in theoretical and experimental research and a discussion of the properties of reverberation from a unified point of view, applying the statistical approach.

§ 2. Sea Reverberation and Its Analogy with Other Physical Phenomena

As sound propagates in the sea it is partially scattered by various inhomogeneities of the medium and irregularities of the boundaries of the latter. The scattering of sound may be elicited in principle by the following objects and irregularities:

Air bubbles;

Fish, marine creatures, microorganisms;

Solid particles;

Temperature irregularities;

Irregularities of the ocean surface;

Irregularities of the bottom and inhomogeneities of the bottom soil composition;

Any other irregularities that might result in abrupt changes or fluctuations in the propagation velocity of sound and density of the water.

Sea reverberation refers to a process describing the time variation of the total scattered sound field observed at the point of reception following transmission of a sound signal. The onset of reverberation is therefore attributable to the transmission of signals and their propagation in a statistically inhomogeneous medium.

In many cases of the investigation of reverberation under oceanic conditions, unfortunately, it proves unfeasible to pinpoint exactly the causes of sound scattering or to classify scatterers according to their type, relative contribution, and characteristics. Reverberation is often attributable to combined scattering by several types of irregularities; hence one cannot ascribe it in every isolated case to some specific type. To a certain extent this complicates the investigation of the several properties of reverberation signals.

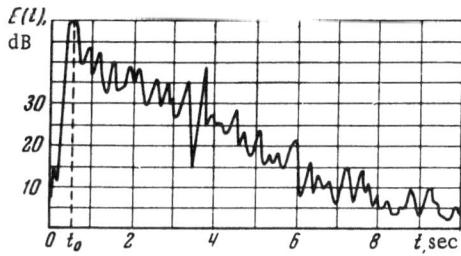

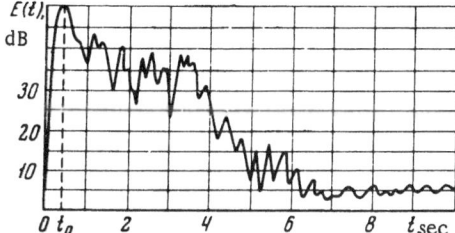

Fig. 1. Time-base plots of reverberation level from a recording instrument of the type N-110. The duration of the transmitted pulses is 300 msec, the carrier frequency is 7 kc; the pulse is transmitted at the time t_0.

Once, however, a definite hypothesis is adopted regarding the distribution of scatterers in the ocean, as well as their possible sizes and acoustical properties, it is possible to carry out a fairly complete analysis of the statistical characteristics of reverberation without analyzing in detail the scattering by all possible types of inhomogeneity. Such a model of reverberation as a stochastic process, portraying it as a sum of elementary scattered signals with stochastic parameters, is discussed below in Sec. 4.

Reverberation appears at the reception point as a fluctuating and gradually time-decaying process. The variation of the mean reverberation level may amount to 30 or 40 dB within the span of a few seconds, in some cases even more. An example of reverberation plots taken from a logarithmic level recorder is shown in Fig. 1; oscillograms of the same, obtained by means of a photographic accessory, are shown in Fig. 2. These records give some idea as to the law of variation of the reverberation level with time and the nature of the fluctuations of its envelope.

Reverberation signals are perceived by the ear as a gradually dying sound (echo tone), the beginning of which corresponds to the instant of pulse transmission.

It is evident, therefore, from a consideration of the elementary properties of reverberation, that it is a nonstationary stochastic process.

There exists a definite analogy between sea reverberation and certain other physical effects. The first to come to mind is reverberation in rooms or reverberation of an enclosure in response to the transmission of signals. Reverberation in rooms is caused by the reflection of sound from the walls and other obstacles, and since multiple reflection is possible in this situation, a summation of elementary signals having random amplitudes and phases is observed at the point of reception.

It is interesting to note that, despite the differences between room reverberation and sea reverberation (dimensions of the medium, causes of decay of the signal levels with time, etc.), the latter is still called reverberation, clearly from the similarity of the physical models of both processes, the possibility of performing measurements with similar acoustical instrumentation, and the likeness of the two processes in their aural perception.

In radar, the process analogous to sea reverberation is called the background due to local objects or chaotic reflections from local objects [2, 7, 8]. This refers to electromagnetic waves reflected and scattered by such objects as vegetation, the surface of the ocean, rain or fog droplets, metal-plated strips, nonuniformities of the ionosphere, etc. The local object background, therefore, like reverberation, is associated with the propagation of waves in a statistically inhomogeneous medium.

Another effect analogous to reverberation is the scattering of light in turbid media.

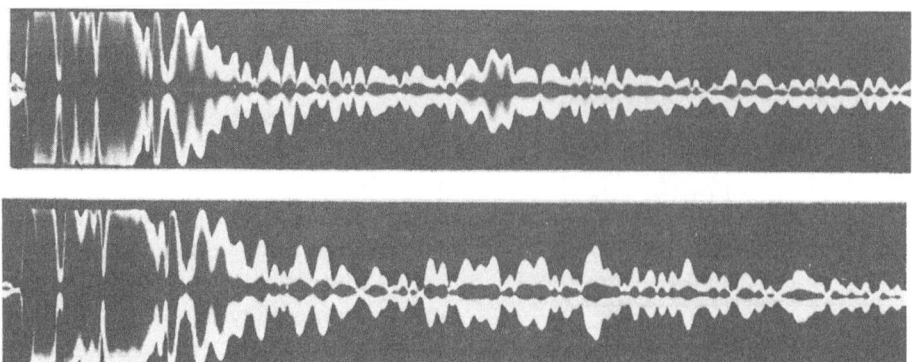

Fig. 2. Reverberation oscillograms obtained by means of a photographic attachment. The duration of the transmitted pulses is 10 msec; the carrier frequency is 25 kc.

Besides those enumerated above, reverberation as a stochastic process is analogous from the point of view of its mathematical description to a whole series of other effects unrelated to wave scattering. We have in mind certain types of fluctuation processes, which, like reverberation, may be thought of as a summation of stochastic perturbations. Among such processes, for example, are thermal noise and certain types of electronic instrument noise, in particular noise due to the shot effect. The development of probabilistic methods to investigate the properties of shot noise has resulted in the formulation of a number of theorems with regard to the superposition of stochastic perturbations [1, 12, 51]; we will make use of these theorems in analyzing the statistical characteristics of reverberation signals.

Consequently, on the one hand, due to the definite analogy that exists between reverberation and other physical phenomena, the study of its statistical properties may be based on the appropriate mathematical apparatus of the theory of stochastic processes that has been developed in application to scattering problems and the analysis of certain types of electrical fluctuations. On the other hand, the features of reverberation as an underwater acoustical phenomenon introduce a distinctive uniqueness in its methods of analysis and imbue its investigation with an inherent nature all its own, thus promoting the development of statistical concepts for different applications.

§ 3. Energy Attributes of Reverberation

The energy approach to the investigation of the properties of reverberation signals permits their mean intensity to be determined as a function of the time and the following characteristics of the transmitted signals, acoustic arrays, and the ocean medium:

Radiated power;

Duration and shape of the signal;

Directionality of the transmitting and receiving arrays;

Sound absorption;

Scattering properties of the ocean medium.

The energy theory of reverberation has been described in sufficient detail in [5, 6, 11, 20-22, 24, 25, 34] and many others; we will discuss only briefly, therefore, without derivations, the results of this theory, focusing attention on the nonstationary quality of the reverberation signals and the values of the scattering coefficients that correspond to various types of inhomogeneities of the ocean medium.

Three types of reverberation are normally distinguished, namely:

Volume, due to scattering by inhomogeneities of the ocean medium proper (such inhomogeneities include, for example, microorganisms, fish, thermal irregularities, etc.);

Surface, caused by the scattering of sound by air bubbles in the surface layer and by ocean surface waves;

Bottom, due to scattering by inhomogeneities of the ocean bottom and the irregularities of its surface.

This classification, however, is not altogether advantageous from the point of view of the mathematical description of the laws governing the decay of the mean reverberation level with time. Specifically, the laws of surface reverberation may prove valid for deep-water sound-scattering layers; the reverberation due to scattering by the ocean surface and bottom is sometimes described by similar relations, etc. It is convenient for this reason to define the following types of reverberation (see, in particular, [21, 24]):

Volume reverberation, due to scattering by inhomogeneities occupying an unbounded space;

Layer reverberation, due to scattering by inhomogeneities concentrated in a layer;

Boundary reverberation, due to scattering by an interface separating two media.

This classification is suitable for the mathematical description of volume, surface, and bottom reverberation, inasmuch as these represent a linear combination of two (and sometimes three) types of the above-defined reverberation processes under certain conditions. For this reason the separate study of volume, surface, and bottom reverberation has a certain practical justification, because it is in just this type of analysis that the laws of decay of the mean level are the most comprehensively investigated and differentiated, that the model of reverberation as a temporal process is verified, and that the sound fields in the ocean medium are related to the reverberation signals picked up by the acoustic receiving array. Moreover, the additive character of the given processes makes it easy to interpret particular results in the analysis of the set of volume, surface, and bottom reverberation without any sacrifice for thoroughness of investigation.

The energy theory of reverberation, as treated in the overall survey of the papers indicated above, is approximate for the following reasons:

This theory does not account for the peculiarities of sound propagation that are associated with reflection in the ocean, namely the mean velocity of sound propagation is regarded as constant in depth.

The scattering properties of the ocean medium are assumed to be statistically uniform in the horizontal plane.

Secondary scattering is disregarded.

In addition, the limits of applicability of the energy theory are determined by the condition

$$t \gg \delta_{ef}, \tag{3.1}$$

where t is the running time, δ_{ef} is the effective duration of the transmitted signals. This implies that the theory describes the so-called far reverberation.

We also mention certain notions dealing with the properties of the scattering inhomogeneities and forming the basis of calculation of the mean reverberation levels; they are also discussed in Sec. 4 in connection with the statistical model of reverberation. It is assumed that the scatterers in the ocean are distributed throughout space within the limits of the scattering region with statistical independence and sufficiently wide spacing that the elementary scattered signals are added incoherently at the point of reception. This immediately permits the assumption that the mean reverberation intensity $<F^2(t)>$* is proportional to the average number of scatterers found in half the scattering region or, what amounts to the same thing, it is proportional to the quantity $<n_1(t)> \delta_{ef}$, where $<n_1(t)>$ is the average number of elementary signals arriving at the reception point per unit time. We note that $<F^2(t)>$ and $<n_1(t)>$ are functions of the time, so that the averaging here is taken over the ensemble of states or over a set of spaces with identical average scattering conditions.

Let us consider the fundamental relations governing the mean reverberation intensity for different distributions of the scatterers in the ocean.

Under the above assumptions we have the following relation for the mean reverberation intensity due to scattering by inhomogeneities occupying an unbounded space:

$$\langle F^2(t) \rangle = \frac{W_A k_0 \delta_{ef} \eta_0}{2\pi c t^2} \cdot 10^{-0.1\beta ct}, \tag{3.2}$$

where $<F^2(t)>$ is the mean intensity in W/cm^2, W_A is the radiated acoustic power in W, c = $1.5 \cdot 10^5$ cm/sec is the sound propagation velocity, t is the time in sec, the parameter β is the coefficient of volume absorption of sound in the sea in dB/km, which is satisfactorily described by the empirical formula

$$\beta = 0.036 f_0^{3/2}, \tag{3.3}$$

f_0 is the mean frequency of the transmitted-signal spectrum in kc,

$$k_o = \frac{dW_s}{\langle F_1^2 \rangle \, dV}, \tag{3.4}$$

k_0 is the volume scattering coefficient in cm^{-1}, which is determined by the scattering power dW_s in watts associated with the volume dV in cm^3 of the scattering space for the incident wave intensity $<F_1^2>$ in W/cm^2†; η_0 is a coefficient depending on the form of the directivity character-

---

*The angle brackets $< >$ always indicate the statistical average of the enclosed stochastic functions over the ensemble of states.

† The coefficient of volume scattering k_0 is numerically equal to the scattering power in W associated with a volume of 1 cm^3 with $<F_1^2> = 1$ W/cm^2. The coefficient k_0 is sometimes used

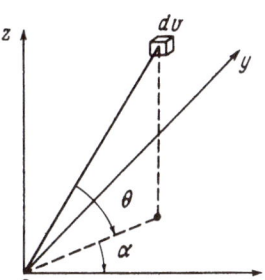

istics of the acoustic arrays and is equal to

$$\eta_0 = \frac{\int\limits_{-\pi/2}^{\pi/2} \int\limits_0^{2\pi} \varphi_T^2(\alpha, \theta)\, \varphi_R^2(\alpha, \theta)\, d\alpha \cos\theta\, d\theta}{\int\limits_{-\pi/2}^{\pi/2} \int\limits_0^{2\pi} \varphi_T^2(\alpha, \theta)\, d\alpha \cos\theta\, d\theta}, \qquad (3.5)$$

where $\varphi_T(\alpha, \theta)$ and $\varphi_R(\alpha, \theta)$ are the normalized directivity characteristics of the transmitting and receiving arrays, respectively.

The coordinate system used for the above relations is shown in Fig. 3 for their interpretation.

Fig. 3. Coordinate system used in calculating the mean reverberation intensity.

The mean reverberation intensity due to scattering by inhomogeneities concentrated in a layer of thickness h is defined as

$$\langle F^2(t) \rangle = \frac{W_A k_0 h \delta \mathrm{ef} \eta_S}{2\pi c^2 t^3} \cdot 10^{-0.1\beta ct}, \qquad (3.6)$$

where η_S is a coefficient related to the directivity of the arrays by the expression

$$\eta_S = \frac{2\int\limits_0^{2\pi} \varphi_T^2(\alpha, \theta_2)\, \varphi_R^2(\alpha, \theta_2)\, d\alpha}{\int\limits_{\pi/2}^{\pi/2} \int\limits_0^{2\pi} \varphi_T^2(\alpha, \theta)\, d\alpha}. \qquad (3.7)$$

Here, as before, k_0 is the volume scattering coefficient in the layer. However, it is often customary to use the so-called coefficient of surface scattering k_S, a dimensionless quantity:

$$k_S = k_0 h. \qquad (3.8)$$

Then, taking (3.8) into account in place of (3.6), we write

$$\langle F^2(t) \rangle = \frac{W_A k_S \delta \mathrm{ef} \eta_S}{2\pi c^2 t^3} \cdot 10^{-0.1\beta ct}. \qquad (3.9)$$

According to (3.4) and (3.8), the quantity k_S determines the scattering power associated with unit area of the layer, i.e.,

$$k_S = \frac{dW_s}{\langle F_1^2 \rangle dS}, \qquad dS = dV/h. \qquad (3.10)$$

to represent the quantity

$$k_0 = \frac{4\pi dW_s}{\langle F_1^2 \rangle dV}, \qquad (3.4.A)$$

so that the coefficient of volume scattering is referred to a unit solid angle. In particular, such a definition of k_0 is used for the analysis of experimental data relating to the scattering properties of the ocean (see, for example, [34, 38]).

The mean reverberation intensity due to scattering by the interface between two media is defined in the form

$$\langle F^2(t) \rangle = \frac{W_A k_b H \delta_{ef} \eta_b}{\pi c^3 t^4} \cdot 10^{-0.1\beta ct}, \qquad (3.11)$$

where H is the distance from the acoustic array to the scattering boundary, k_b is the scattering coefficient of the interfacial boundary, and is equal to

$$k_b = \frac{\langle F_2^2 \rangle}{\langle F_1^2 \rangle} = \frac{dW_s}{\langle F_1^2 \rangle \, dS}, \qquad (3.12)$$

where $\langle F_1^2 \rangle$ is the mean intensity of the incident wave and $\langle F_2^2 \rangle$ is the mean intensity of the scattered wave, measured at unit distance from an elementary area of the interface; η_b is a coefficient coinciding with (3.7).

It is clear that the relations (3.2), (3.9), and (3.11) may be written as a function of the distance, since the time t is related to the distance R to the scattering region by the expression

$$t = 2R/c.$$

It follows from the above relations that the following general principles are typical of the mean reverberation intensity:

The mean intensity is proportional to the radiated acoustic power, effective signal duration, and a coefficient depending on the directivity of the acoustic arrays.

The mean intensity diminishes with increasing sound absorption in the sea.

The decay of the mean intensity with time is characterized by a power law with exponents 2, 3, and 4 for reverberation due to scattering by inhomogeneities situated in an unbounded medium, in a layer, and at an interface between two media, respectively.

Under real conditions, as a rule, the coexistence of several types of reverberation signals is observed. In the general case we may write for the mean reverberation intensity, on the basis of the relations (3.2), (3.9), and (3.11),

$$\langle F^2(t) \rangle = A_1(t) \frac{W_A k_0 \delta_{ef} \eta_0}{2\pi c t^2} \cdot 10^{-0.1\beta ct} + A_2(t) \frac{W_A k_S \delta_{ef} \eta_S}{2\pi c^2 t^3} \cdot 10^{-0.1\beta ct} + A_3(t) \frac{W_A k_b H \delta_{ef} \eta_b}{\pi c^3 t^4} \cdot 10^{-0.1\beta ct}, \qquad (3.13)$$

where $A_1(t)$, $A_2(t)$, and $A_3(t)$ are certain functions defining the geometric position of the scattering regions in space, the anomalies of sound propagation, and the orientation of the directivity characteristics of the acoustic arrays.

Numerous experiments have been performed for confirmation of the theoretical laws governing the decay of the mean reverberation intensity as a function of time. Figure 4 shows as an example analytical curves and experimental data borrowed from [24] for $\langle F^2(t) \rangle$ for the three types of reverberation, volume, surface, and bottom. It is evident from the graph that the experimental data agree with the theoretical principles.

It turns out, in particular, that Eq. (3.2) well describes the time decay law for volume reverberation, Eq. (3.9) for surface reverberation, and Eq. (3.11) for bottom reverberation. Here, k_0, k_S, and k_b are interpreted as the coefficients of volume, surface, and bottom scattering, respectively.

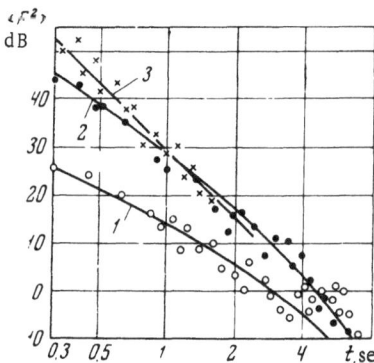

Fig. 4. Comparison of the theoretical and measured laws for the decay of the reverberation level with time according to the data of [24]. 1) For volume reverberation; 2) for surface reverberation; 3) for bottom reverberation.

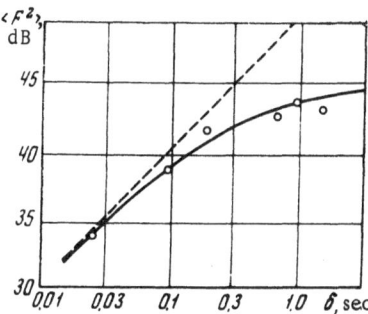

Fig. 5. Dependence of the reverberation level on the signal duration according to the data of [24].

In [24, 25] the dependences of the reverberation level on the duration of the transmitted signal are discussed. These dependences are shown in Fig. 5 according to the data of [24] for rectangular signals. The linear relation between $<F^2>$ and δ proves valid over a wide range of variation of signal durations. It should be remembered, however, that such a dependence is only observed up to definite values of δ, that the reverberation level then depends more weakly on the duration, eventually becoming almost independent of it. This result is reasonable, because, for large pulse durations, the condition (3.1), for which the given energy theory is valid, is not satisfied.

[It is pointed out in the papers cited above that the linear dependence of $<F^2>$ on δ is violated for small times t and for large absorption of sound on the interval $(t, t + \delta)$.]

For an estimate of the reverberation level under different oceanic conditions, one often makes use of the concepts of reverberation force and the so-called equivalent radii (see, e.g., [25]).

By equivalent radius r_e we mean the radius of a specularly reflecting sphere which, when placed at a distance equal to the distance from the scattering region, will create at the reception point an echo signal whose intensity coincides with the mean reverberation intensity.

Under the given assumptions, the mean intensity of a signal reflected from a sphere situated on the axis of the transmitter directivity characteristic may be defined as

$$\langle F_c^2(t) \rangle = \frac{W_A \eta_c r_e^2}{4\pi c^4 t^4} \cdot 10^{-0.13ct} , \qquad (3.14)$$

where η_c is a parameter depending on the directivity of the transmitter and equal in this case to its axial concentration coefficient:

$$\eta_c = \frac{4\pi}{\int\limits_{-\pi/2}^{\pi/2} \int\limits_0^{2\pi} \varphi_T^2(\alpha, \theta)\, d\alpha \cos\theta d\theta} . \qquad (3.15)$$

On the basis of Eqs. (3.2), (3.9), (3.11), and (3.14) we obtain the following for the equivalent radii of volume, surface, and bottom reverberation, respectively, equating the intensities $<F^2(t)>$ and $<F_c^2(t)>$:

$$r_{e0} = (2c^3 k_0 \delta_{ef} \eta_0/\eta_c)^{1/2} t, \qquad (3.16)$$

$$r_{eS} = c \, (2k_S \delta_{ef} \eta_S / \eta_c)^{1/2} \, t^{1/2},$$
(3.17)

$$r_{eb} = 2 \, (ck_b H \delta_{ef} \eta_b / \eta_c)^{1/2},$$
(3.18)

where the parameters η_0 / η_c, η_S / η_c, and η_b / η_c, proceeding from (3.5), (3.7), and (3.15), are determined by the relations

$$\eta_0 / \eta_c = (1/4\pi) \int_{-\pi/2}^{\pi/2} \int_0^{2\pi} \varphi_T^2 (\alpha, \theta) \, \varphi_R^2 (\alpha, \theta) \, d\alpha \cos\theta d\theta,$$
(3.19)

$$\eta_s / \eta_c = \eta_b / \eta_c = (1/4\pi) \int_0^{2\pi} \varphi_T^2 (\alpha, \theta_0) \, \varphi_R^2 (\alpha, \theta_0) \, d\alpha.$$
(3.20)

The concept of equivalent reverberation radii proves particularly suitable for the mutual comparison of different types of reverberation levels, as well as for their comparison with the levels of reflection from different obstacles, for assessing the influence of the characteristics of sound propagation on the intensity of underwater sound signals, etc.

It is apparent from (3.16)-(3.18) that for given parameters of the transmitted signals and directivity characteristics of the acoustic arrays the equivalent radii of reverberation will depend on its type, the time, and the scattering coefficients of the ocean medium.

We look now at the experimental data relating to the scattering coefficients of the ocean and its boundaries. We note that the values of these coefficients are often expressed in decibels and are denoted, respectively,

$$\left. \begin{array}{l} m_0 = 10 \log k_0, \\ m_S = 10 \log k_S, \\ m_b = 10 \log k_b, \end{array} \right\}$$
(3.21)

where m_0 has the conditional dimensions dB/cm (or dB/m), while m_S and m_b have the dimensions dB.

The values of the scattering coefficients depend in general on many factors, the principal of which are the following:

Type of scatterers;

Carrier frequency of the transmitted signals;

Angle of incidence on the scattering region;

State of the ocean surface;

Type of bottom;

Time of year and time of day.

There clearly exist other factors that would cause the experimental values of the scattering coefficients to have a significant spread, deviating in some cases by several tens of decibels. This leads in particular to difficulties in predicting the mean reverberation levels in certain specific situations and makes it troublesome to use averaged values in working with hydroacoustic instruments in poorly studied regions.

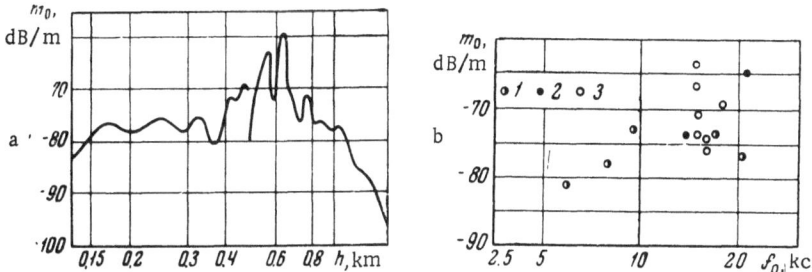

Fig. 6. Experimental values of the volume scattering coefficients according to the data of [38]. a) Dependence on depth; b) dependence on frequency for three different regions (1, 2, 3).

The acquisition of experimental data on the scattering coefficients of the ocean is always dependent on the characteristics of the ocean measurements; a factor of extreme importance here is the monitoring of the sound propagation conditions. This is particularly significant, because the relations for $<F^2>$ or r_e given above, which are normally used to determine the scattering coefficient, are formulated for the ocean medium without regard for acoustic refraction. In this connection any propagation anomalies would necessarily affect the values of the coefficients to be measured and must therefore be carefully monitored.

It is interesting to note in this connection the method of measuring the so-called local scattering coefficients by the transmission of short pulses or using explosive sound sources, for which relatively small volumes of the ocean are subjected to investigation. This method is particularly fruitful in the investigation of the volume and surface scattering coefficients but it is also used to study the coefficients of bottom scattering; among its advantages is the capability it affords of performing detailed measurements of the distribution of scattering levels with respect to the various spatial coordinates and the absence of any influence from the refraction of sound.

At the present time there are a great many published works containing materials on the investigation of scattering levels. One group of papers is concerned with the theoretical analysis of the properties of the ocean medium; in these papers scattering by air bubbles, temperature irregularities, microorganisms, fish, ocean surface and bottom roughness, and other types of scatterers. A second group of papers is devoted to the experimental investigation of the scattering coefficients of the water medium and its boundaries.

A brief survey is given below of the experimental data on the values of the coefficients of volume, surface, and bottom scattering.

According to the results of most researchers (see, in particular, [24, 25, 34, 38, 39, 47]), the values of the coefficient m_0 lie between the limits from −40 to −100 dB/m. For example, in [34], values from −50.4 to −90.4 dB/m are given for m_0, in [38] the experimental values of m_0 were determined in the interval from −64 to −81 dB/m. Also given in the latter paper are the results of measurements of m_0 as a function of depth and frequency. The measurements were performed in three different regions of the southwestern Atlantic Ocean. These data are presented in Fig. 6. It follows from the graph for the depth dependence of the volume scattering coefficient that at certain levels an increase in the sound scattering is observed, the inhomogeneities in these cases being more or less locally disposed in the body of the water.

The effects of increased scattering in layers lying at depths from 100 to 1000 m (deep sound-scattering layers) are attributed, according to the data of most studies, to the higher

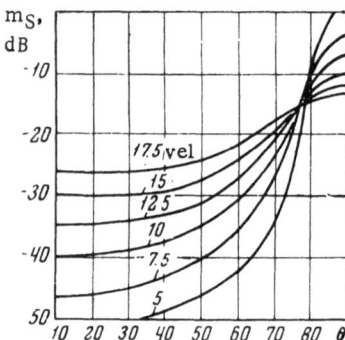

Fig. 7. Dependence of the surface scattering coefficients on the grazing angles for various wind velocities according to the data of [57].

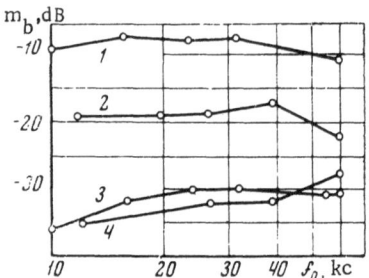

Fig. 8. Dependence of the coefficient of bottom scattering on the frequency according to the data of [56]. 1) Sand and pebbles; 2) slime and mussels; 3, 4) slime.

concentration in them of living organisms, mainly microorganisms and fish. These layers migrate, and their scattering properties in general vary during the course of the day, depending also on the time of year and geographic locale. Furthermore, thermal irregularities, high temperature gradients, and variations of the water density are cited as possible reasons for the increase in volume scattering level; also discussed by way of conjecture is the possibility of a concentration near depths with abrupt changes of temperature and density of the water of living organisms, falling waste matter, and other inhomogeneities, which could give rise to enhanced sound scattering under certain conditions.

Investigations of the acoustical characteristics of deep sound-scattering layers have been performed by a variety of methods — by the transmission of pulses, tone signals, and by means of explosive sources. In certain papers (see, e.g., [38, 47]), the presence of peaks in the volume scattering levels is noted, corresponding to definite frequencies in the range from 4 to 6 kc. It is remarked in [39] that these peaks often occur at higher frequencies (exceeding 5 kc). The position of the peaks on the frequency scale varies during the course of the day, a fact which, in the opinion of the authors of these papers, suggests a biological origin of the investigated sound-scattering layers.

The results of investigations of the energy properties of surface reverberation are given in numerous theoretical and experimental papers (see [20, 24, 25, 33, 34, 36, 46, 53, 57] and others). It is mentioned in particular that the values of the surface scattering coefficients m_S depend on the angles of incidence of the sound on the scattering region, the state of the ocean surface, and a number of other factors. According to the data of the majority of authors, the value of m_S lies within the interval from −10 to −60 dB.

In [57], for example, typical dependence of the values of m_S on the grazing angle are given for various wind velocities. These dependences are shown in Fig. 7. It is noted in the cited reference that at grazing angles in the interval from 0 to 60-70° surface reverberation is due primarily to sound scattering by air bubbles in the surface layer; for grazing angles in excess of 70-80° reverberation seems to be associated with scattering by the roughness of the ocean surface itself. The dependence of the reverberation level on the wind force also indicates that air bubbles are the scatterers in this case and also determined the measured values of m_S. Data obtained by other authors on the values of m_S generally agree with those presented above.

A considerable number of investigations are also concerned with bottom reverberation levels. The coefficients of bottom scattering m_b, according to the data of most papers [24, 25, 34, 44, 48, 56], lie within the interval from −5 to −35 dB and depend strongly on the type of bottom soil and grazing angles. According to these papers (with the exception of [48]), no kind of system-

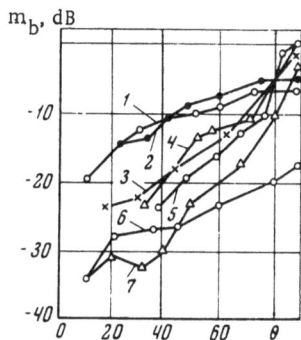

m_b, dB

Fig. 9. Dependence of the coefficients of bottom scattering on the grazing angles for frequencies of 55 and 60 kc according to the data of [56]. 1, 2) Sand and rock, 55 kc; 3) sand, 60 kc; 4, 6) slime, 60 kc; 5) slimey sand, 60 kc; 7) slime, 55 kc.

atic frequency dependence is exhibited by m_b over a wide range (from a few kilocycles to several hundred kilocycles). In particular, the values of the bottom scattering coefficients m_b are given in [25, 56] for various frequencies and types of bottom soil (sand, rocks, slime, mussels, etc.). The results of measurements of m_b for various frequencies are shown in Fig. 8 as an example.

We also mention [44], which contains the results of measurements of the scattering coefficients of an ocean bottom of the red slime type, which covers 26% of the bottom of the Atlantic type and 49% of the Pacific. The measurements were made at frequencies of 530 and 1030 cps, the values of m_b turning out to be equal to -27 dB at both frequencies.

In conjunction with the results of [56] these data lead one to regard the values of m_b as practically independent of the frequency in the range of the above families of events for the indicated type of bottom, as well as for slime and sand bottoms.

An interesting article in this respect is [45], in which attention is focused on the frequency independence of the levels of back-scattering by surface roughnesses for electromagnetic, elastic, and acoustic waves. This independence, which is observed over a very wide range of frequencies, clearly indicates the existence of scattering inhomogeneities with the most widely diverse scales.

The lack of a frequency dependence on the part of m_b is not, however, a universal property of bottom scattering. In one of the later papers [48], in particular, it is indicated that for certain types of ocean bottoms the resulting experimental dependence of the bottom scattering coefficient on the frequency has the form

$$m_b \sim 10 \log f^k,$$

where the exponent k is determined by the type of bottom. For sand and gravel bottoms, for example, the measured value of k turned out to be near 1.6; for firm bottoms m_b does not disclose a frequency dependence.

In some of the papers indicated above data are also presented on the dependence of the bottom scattering coefficients on the grazing angle. It is remarked in certain studies, in particular, that the scattering coefficient increases as a rule with angle of sound incidence, reaching a maximum at normal incidence.

In [44, 48] and a number of others, for example, the results are given from investigations, exhibiting an increase in the scattering level with increasing angle θ according to a $\sin^p\theta$ law, where the value of p varies between 0 and 2, depending on the size of the angle and type of bottom. The following empirical formula is cited in this connection as valid for the coefficient of bottom scattering m_b:

$$m_b = 10 \log k_{b_1} + 10 \log \sin^p \theta, \qquad (3.22)$$

where $m_{b_1} = 10 \log k_{b_1}$ is the bottom scattering coefficient for $\theta = 90°$.

Sample experimental data taken from [56] on the value of m_b as a function of the grazing angle are shown in Fig. 9.

In many of the papers cited above an analogy is also made between the dependences of bottom and surface scattering on the grazing angle, which in some cases proves helpful. An interesting paper in this connection is [50], in which certain patterns in the scattering of sound, light, and electromagnetic waves by statistically rough surfaces are compared, and a possible mechanism to explain, in particular, the analogous dependences of the scattering coefficients on the grazing angle is analyzed, these dependences approaching the law (3.22) over a definite range of angles.

We conclude this section with mention of the fact that it is sometimes possible to observe so-called coherent scattering, which is caused by the presence in the ocean of inhomogeneities of the air bubble or other discrete types, separated from one another by distances less than the acoustic wavelength. Certain aspects of coherent scattering have been treated in [5, 6, 11], where several of the laws governing this type of scattering are analyzed. Coherent scattering can affect the energy and statistical characteristics of reverberation and create a certain regular component in the reverberation signals. The onset of coherent scattering is associated with the presence of clustered inhomogeneities with rather sharply delineated boundaries or appreciable variations in the mean scatterer density throughout space. In particular, this type of scattering may be observed upon irradiation of a dense, sharply delineated school of fish or a localized cluster of air bubbles with a high concentration, etc.

Unfortunately, experimental data on the levels of coherent scattering are lacking at the present time.

§4. Model of Reverberation as a Stochastic Process

It is necessary for the analysis of the statistical properties of reverberation to adopt some physical model, which must reflect the probabilistic characteristics of the scattering process and take into account the acoustical properties of the scatterers. In Sec. 3, where we presented the energy theory of reverberation, certain postulates have already been stated with regard to such a model. In particular, the introduction of the volume scattering coefficient k_0, according to the definition (3.4), and the assumption that the acoustic power scattered by a certain volume is proportional to the size of that volume actually correspond to the incoherent addition of elementary scattered signals for k_0 = const.

However, a reverberation model consistent with the energy theory does not adapt to the investigation of its statistical properties, because of its incompleteness. We need therefore another, more complete model, by means of which it would be possible, on the one hand, to study the laws intrinsic to reverberation as a statistical phenomenon and, on the other, to derive its energy characteristics as a special result.

We proceed now with a description of the statistical model of reverberation.

We assume that the scatterers in the ocean are located discretely and we denote by a_i and t_i the stochastic amplitudes and onset time of the elementary scattered signals. This is an important theoretical postulate, and its legitimacy will be discussed below.

If the transmitted signal is described by some function of the time s(t) and every elementary scattered signal reproduces these functions in form, then any state of the reverberation process $F_n(t)$ at the reception point may be described at a fixed time t in the form

$$F_n(t) = \sum_{i=1}^{n} a_i s(t - t_i). \tag{4.1}$$

The summation in (4.1) is taken over all possible values of the stochastic variables a_i and t_i corresponding to the scattering region in space. The number of elementary signals n producing reverberation at the reception point at the time t is also a stochastic variable.

The representation (4.1), as already mentioned, is based on the discrete nature of the scattering properties of the ocean.

We now take a further look at the justification for such a reverberation model, particularly the problem of whether the observed reverberation signals are indeed at least represent-able in the form of a sum of the type (4.1).

The effective scattering cross sections of such scatterers as air bubbles, fish, microorganisms, solid particles, inhomogeneities of many types of bottom soil, etc., turn out in practice to be considerably smaller than the distances occupied by the transmitted signals in space, and the scattering cross sections of some of them are less than the acoustic wavelength. From the viewpoint of the possible dimensions of scatterers, then, the representation (4.1) is valid with an accuracy sufficient for practical purposes. A possible exception in this sense is scattering by thermal irregularities of the ocean medium, by a statistically rough ocean surface, or by large-scale roughness of the ocean bottom. It is essential in this case, if at all possible, either to divide the scattering volume into effective scatterers with dimensions of the same order as the spatial correlation region of the inhomogeneities, or to treat the influence of large scatterers separately.

Moreover, even if the properties of the ocean inhomogeneities vary smoothly as a function of the spatial coordinates, the total scattered signal in this case may be represented in principle as a sum, just as any random process is represented in the form of a canonical expansion [19]. Then the following question arises. Will the canonical expansion of the function F(t) be equivalent to a representation of the type (4.1)? In some cases a definite lack of correspondence will occur, because the terms of the expansion will not necessarily be described by the same type of functions, as implied by (4.1). For example, one may observe "elongation" of the elementary scattered signals relative to the transmitted signal, changes of shape or duration, etc. In this case it is convenient to use a more general representation of the type (4.5) for the reverberation process, as discussed below.

Consequently, the assumption of a discrete character on the part of reverberation does not essentially violate the generality of the investigation of its properties; as so aptly brought out in [35], the case of a continuous scattering medium is obtained at once by increasing the number of scattered signals generating the reverberation process, and in this sense the discrete scattering model is innate, not a fictitious assumption.

If the average number of scatterers per unit volume of any part of space is constant, then, for a given instant of time t there exists a certain value of $<n_1(t)>$ equal to the average number of elementary scattered signals arriving at the reception point per unit time.

The parameter $<n_1(t)>$ for the reverberation process generally depends on the running time, because, in the propagation of sound, the scattering volume varies due to divergence of the wave front and, in addition, the scattering properties of the ocean can vary throughout space. If, however, the effective duration of the transmitted signals is equal to δ_{ef}, then given the conditions (3.1), it is reasonable to assume on the time interval $(t - \delta_{ef}/2, t + \delta_{ef}/2)$ the parameter $<n_1(t)>$ is constant and equal to $<n_1>$.

We will also consider that the Poisson distribution law holds for the number n of elementary scattered signals arriving at the reception point during the time $T \ll t$, i.e., in the time interval $(t - T/2, t + T/2)$:

$$P(n) = \frac{(\langle n_1 \rangle T)^n}{n!} \exp(-\langle n_1 \rangle T). \tag{4.2}$$

The distribution law (4.2) proves valid for the following two assumptions relative to the scattering properties of the ocean medium:

The scatterers in the medium are statistically independent in position.

For some sufficiently large scattering region D of the medium the mean scatterer density is assumed to be constant throughout space.

The first assumption is necessary for justification of the distribution law $P_N(n)$ governing the number n of scatterers falling into some region of the medium $d \in D$ if the region D contains N scatterers.

This law is binomial:

$$P_N(n) = C_N^n P^n (1-P)^{N-n}, \tag{4.3}$$

where the probability P is defined as

$$P = d/D.$$

The second condition, which must be written in the form

$$\lim_{N \to \infty} PN = \langle n_1 \rangle T,$$

introduces a mathematical restriction necessary for the limiting transition as $N \to \infty$ from the distribution (4.3) to the Poisson law (4.2).

We note that for a finite but sufficiently large value of N the distribution law for the number of scatterers is almost a Poisson law, so that it is not necessary in principle to make the limiting transition as $N \to \infty$ in order for that law to be valid.

In fact, the binomial distribution (4.3) may be written in the form

$$P_N(n) = \frac{(PN)^n}{n!} \left(1 - \frac{PN}{N}\right)^N \left(1 - \frac{PN}{N}\right)^{-n} \left(1 - \frac{1}{N}\right)\left(1 - \frac{2}{N}\right) \cdots \left(1 - \frac{n-1}{N}\right).$$

The evaluation of the cofactors of this expression for N > 1 yield, respectively,

$$\left(1 - \frac{PN}{N}\right)^N = \exp(-PN)\left[1 - \frac{(PN)^2}{2N} + 0\left(\frac{1}{N^2}\right)\right],$$

$$\left(1 - \frac{PN}{N}\right)^{-n} = 1 + \frac{nPN}{N} + 0\left(\frac{1}{N^2}\right),$$

$$\left(1 - \frac{1}{N}\right)\left(1 - \frac{2}{N}\right) \cdots \left(1 - \frac{n-1}{N}\right) = 1 - \frac{n(n-1)}{2N} + 0\left(\frac{1}{N^2}\right).$$

Consequently, the distribution $P_N(n)$ may be written in the form

$$P_N(n) = \frac{(PN)^n}{n!} \exp(-PN)\left\{1 - \frac{1}{N}\left[\frac{n^2}{2} - n\left(PN + \frac{1}{2}\right) + \frac{(PN)^2}{2}\right] + 0\left(\frac{1}{N^2}\right)\right\}.$$

If we assume that N is sufficiently large, the following approximate equation is valid:

$$PN \approx \langle n_1 \rangle T.$$

Then the binomial distribution $P_N(n)$ is represented as the product of the Poisson law (4.2) and a series that converges rapidly to unity, i.e.,

$$P_N(n) = \frac{(\langle n_1 \rangle T)^n}{n!} \exp\left(-\langle n_1 \rangle T\right) \left\{ 1 - \frac{1}{N} \left[\frac{n^2}{2} - n\left(\langle n_1 \rangle T + \frac{1}{2}\right) \right] + 0\left(\frac{1}{N^2}\right) \right\}. \tag{4.4}$$

Hence, it is apparent that for sufficiently large N the distribution (44) is close to a Poisson distribution.

We also estimate the convergence of $P_N(n)$ to $P(n)$, i.e., to the law (4.2).

The maximum value of the expression in the brackets on the right-hand side of (4.4) is obtained for

$$n = \langle n_1 \rangle T + \frac{1}{2}.$$

The following estimate is therefore valid for the difference $|P_N(n) - P(n)|$:

$$P_N(n) - P(n)| \leqslant \left| \frac{4\langle n_1 \rangle T - 1}{8N} \right| + 0\left(\frac{1}{N^2}\right).$$

For example, if we limit the approximation to a Poisson distribution to at least 10%, we obtain the necessary condition

$$N \geqslant 5\left(\langle n_1 \rangle T - \frac{1}{4}\right).$$

If the average number $<n_1>$ of scattered signals arriving simultaneously at the reception point depends on the time, then for the Poisson law in the form (4.2) to be justified, it is necessary that the scattering region be divisible into a certain set of regions $D_1, D_2, \ldots, D_k$, in each of which the Poisson law is valid for the number of scatterers $n_1, n_2, \ldots, n_k$, respectively.

For the total number of elementary scattered signals n in this case the law (4.2) is also valid with a variable parameter $<n_1(t)>$ depending on the time.

We assume further that for the stochastic amplitudes a_i there exists some probability density $W(a)$ and moments $<a^k>$ for this distribution

$$\langle a^k \rangle = \int_0^\infty a^k W(a)\, da$$

of any order k.

If, however, one or more scattered signals have amplitudes that prevail over the amplitudes of the rest, the influence of such signals may be accounted for separately.

We point out that the probability density $W(a)$ depends in general on the running time due to divergence of the wave front, absorption of sound in the sea, and the influence of propagation anomalies.

The form in which the stochastic process is specified in (4.1), as noted above, is a special case of the so-called canonical expansions of stochastic functions, where a_i and $s(t - t_i)$

correspond in the terminology of canonical expansions [19] to the c o e f f i c i e n t s and c o o r d i -
n a t e f u n c t i o n s of the expansion.

The theory of canonical expansions and their application for the analysis of stochastic
processes enable one to determine the fundamental statistical characteristics of the total process F(t) if the properties of the stochastic variables a_i, t_i, and n are known, this analysis being more detailed the more completely the properties of the coefficients and coordinate functions of the expansion are known.

The representation (4.1) corresponds to a model of reverberation in the event that the
elementary scattered signals depend only on the two stochastic parameters a_i and t_i, and their
form coincides with the form of the transmitted signals.

A more general representation of the state of the reverberation process $F_n(t)$ than (4.1)
has the form

$$F_n(t) = \sum_{i=1}^{n} a_i s(t - t_i, \xi_i), \qquad (4.5)$$

where ξ_i is the set of stochastic parameters defining the characteristics of the elementary
scattered signals which may, for example, depend on the displacement of the acoustic arrays,
the motion of the scatterers, their physical properties, their spatial distribution, etc. Such a
reverberation model embraces many practically important cases of sound scattering by various
inhomogeneities of the ocean, and in the ensuing discussion we will use the expression, as well
as (4.1).

Another possible approach to the investigation of the statistical properties of reverberation is the wave version, which is unquestionably more general and more sound physically. This
approach is used [26] to solve many problems in wave scattering during propagation in a medium with random inhomogeneities, the origin of which stems from the presence, for example,
of thermal fluctuations.

The statistical properties of reverberation in this case may be found from the wave equation, which must be solved for specified space−time statistical characteristics of the scattering properties of the ocean medium. This method is used specifically in [29], in which an
analysis is made of the time correlation and spectral characteristics of volume reverberation
for an investigation of the latter as a first-order scattering effect. In the cited paper the solution of the following wave equation is studied:

$$\nabla^2 p - (1/c^2)(d^2 p/dt^2) = 0, \qquad (4.6)$$

in which the velocity of sound propagation c is given in the form

$$c = c_0/(1 + \mu), \qquad \langle \mu \rangle = 0, \qquad [\langle \mu^2 \rangle]^{1/2} \ll 1,$$

where $\mu = \Delta c / c_0$ represents the fluctuations of the refractive index, which depend on the spatial coordinates.

In principle, for the analysis of the statistical characteristics of reverberation, as already stated, it is necessary to find a solution to Eq. (4.6), specifying the space−time statistical characteristics of the parameter μ. However, this solution proves very complex in the
general case, and only special examples of spatial distributions of μ, which do not cover very
many cases of practical significance, may be treated with any measure of success.

But, if we assume that there are sufficiently sharp discontinuities in the refraction coefficient of the medium as a function of the spatial coordinates, it is then possible to go from a consideration of the wave equation to the reverberation model described by the sums (4.1) or (4.5).

In fact, from the solution of the wave equation, we obtain the familiar representation for the process F(t) characterizing the total scattered field at the point of reception; for this process, if we take into account first-order scattering, an integral representation of the following type is valid:

$$F(t) = (2kP_0/4\pi) \int_V \mu(x, y, z) \frac{s(t - 2r/c_0)}{r^2} dV, \quad dV = dx\,dy\,dz, \tag{4.7}$$

where $k = \omega_0/c_0$ is the wave number, the parameter P_0 characterizes the amplitude of the incident wave and accounts for the dimensionality, $\mu(x, y, z)$ are small fluctuations of the refractive index, the function $s(t - 2r/c_0)$ describes the form of the transmitted signals, r is the distance from a scattering element with coordinates (x, y, z) to a point of observation with coordinates (0, 0, 0), so that $r = [x^2 + y^2 + z^2]^{1/2}$. The integration in (4.7) is taken over the entire scattering volume V.

If we now assume that the secondary sources represent a set of scatterers situated sufficiently locally in space, the function $\mu(x, y, z)$ may be represented in the form

$$\mu(x, y, z) = \sum_{i=1}^{\infty} \mu_i \delta(x - x_i, \ y - y_i, \ z - z_i), \tag{4.8}$$

where $\mu_i = \mu(x_i, y_i, z_i)$ is the magnitude of the discontinuity of the refractive index at the site of the i-th scatterer, the delta function $\delta(x - x_i, \ y - y_i, \ z - z_i)$ characterizes the spatial discreteness of the scattering properties of the ocean. The introduction of the delta function means, of course, a substantial idealization of the scattering properties of the ocean. We are not so much concerned here, however, with the physical realization of the representation (4.8) as in making use, as will be apparent from subsequent considerations, of the well-known properties

$$\int_{-\infty}^{\infty} \delta(y)\,dy = 1, \quad X(y_0) = \int_{-\infty}^{\infty} X(y)\,\delta(y - y_0)\,dy,$$

where the function $\delta(y - y_0)$ falls off much more rapidly than the function X(y) in the neighborhoods of the point y_0 for $y > y_0$ and $y < y_0$. Hence, the representation (4.8), being a mathematical idealization of course, merely indicates the local character of the scatterers in the ocean medium, i.e., reduces the reverberation model to a discrete one.

If we also substitute (4.8) into (4.7), for F(t) we have

$$F(t) = (2kP_0/4\pi) \int_V \frac{s(t - 2r/c_0)}{r^2} \sum_{i=1}^{\infty} \mu_i \delta(x - x_i, \ y - y_i, \ z - z_i)\,dV. \tag{4.9}$$

Carrying out the integration of the sum of delta functions in (4.9) and introducing the notation

$$a_i = 2kP_0\mu_i/4\pi r_i^2, \qquad t_i = 2r_i/c_0,$$

we obtain

$$F(t) = \sum_{i=1}^{\infty} a_i s(t - t_i). \tag{4.10}$$

It is apparent that every state $F_n(t)$ of the process F(t), represented in the form (4.10), will be described by a sum of the type (4.1); in other words, we have arrived, starting with the solution of the wave equation, at a discrete reverberation model.

The further analysis of the statistical properties of reverberation will be carried out on the basis of a discrete model of the scattering of sound by inhomogeneities of the ocean medium, on the assumption that the reverberation is a result of the addition of elementary scattered signals with random parameters at the reception point.

§ 5. One-Dimensional Theorem of the Superposition of Stochastic Perturbations

Under the assumptions that were adopted in Sec. 4, reverberation is characterized as a stochastic process resulting from the superposition of a certain number of elementary scattered signals (perturbations) at the point of reception.

Such a reverberation model has distinct analogies, from the viewpoint of its mathematical description, with a great many physical effects. In particular, this model is to a large extent similar to the model of stochastic processes characterizing the scattering of radio waves by surface inhomogeneities of the land and sea, by raindrops, and by other scatterers, as well as to the model of electronic tube noise due to the shot effect.

The indicated analogy makes it possible to use the mathematical theory based on the canonical expansions of stochastic functions for the description of reverberation as a stochastic process. For one of the special forms of canonical expansions used below there exist the so-called theorems of superposition of stochastic perturbations, which have been analyzed in a number of papers [1, 12, 19, 51] in application to the analysis of the properties of various kinds of electrical fluctuations.

The basic assumptions that are made in proving the theorem of superposition of stochastic perturbations with respect to the investigated stochastic process F(t) consist in the following. Any of the states $F_n(t)$, the set of which determines the investigated process F(t), may be represented in the form

$$F_n(t) = \sum_{i=1}^{n} a_i s(t - t_i), \tag{5.1}$$

where a_i and t_i are the stochastic amplitude and onset time of the i-th perturbation, n is a random number of perturbations adding together at the time t, s(t) is a function describing their form. This representation of the stochastic process coincides with the representation (4.1) for the assumed reverberation model.

It is also supposed that the probability P(n) of the number of perturbations n occurring on the interval $(t - T/2, t + T/2)$ obeys a Poisson law:

$$P(n) = \frac{(\langle n_1 \rangle T)^n}{n!} \exp(- \langle n_1 \rangle T) \tag{5.2}$$

and that there exist initial moments $<a^k>$ of the perturbation amplitude distribution $W(a)$ that are obtained by statistical averaging of the values of a_1 over the ensemble of states, i.e.,

$$\langle a^k \rangle = \int_0^\infty a^k W(a)\, da. \tag{5.3}$$

Generally speaking, the parameter $<n_1>$ can depend on the running time, which means that the probability $P(n)$ will also depend on the time. This exhibits, in particular, the nonstationarity of the properties of reverberation signals. Here, however, we will not regard this process in detail, merely indicating the method of calculating the statistical characteristics of the process $F(t)$, assuming it to be stationary.

For the proof of the one-dimensional theorem we consider that all perturbations $s(t - t_i)$ have the same form. This is not really necessary, because in the more general case the process $F_n(t)$ may be represented in the form

$$F_n(t) = \sum_{i=1}^n a_i s(t - t_i,\, \xi_i),$$

which is analogous to (4.5), and it may be assumed that the perturbations have different forms. The function $s(t - t_i, \xi_i)$ in such event can depend on the set of stochastic parameters ξ_i, for example, the carrier frequency, duration of the perturbations, their form, etc. The generalization of the theorems of superposition of stochastic perturbations to processes of the type (4.5) is discussed below in Sec. 7.

We turn now to a consideration of the one-dimensional theorem of superposition of stochastic perturbations, which enables us to compute the semi-invariants of the distribution $W(F)$ of the investigated process (5.1). According to this theorem, the k-th order semi-invariant λ_k is equal to

$$\lambda_k = \langle n_1 \rangle \langle a^k \rangle \int_{-\infty}^\infty s^k(t)\, dt, \tag{5.4}$$

where the normalization of the function $s(t)$, as stated, determines the form of the perturbations.

Knowledge of the semi-invariants is sufficient for computation of the moments α_k and μ_k about the origin and about the mean for the one-dimensional distribution. Hence, by definition,

$$\alpha_k = \int_{-\infty}^\infty F^k W(F)\, dF, \tag{5.5}$$

and

$$\mu_k = \int_{-\infty}^\infty (F - \langle F \rangle)^k W(F)\, dF. \tag{5.6}$$

Inasmuch as the semi-invariants are coefficients of the terms of a power series into which the logarithm of the characteristic function $\Theta(\eta)$ of the probability density $W(F)$ is expanded, the following equation holds:

$$\ln \Theta (\eta) = \ln \langle \exp (j\eta F)\rangle = \sum_{k=1}^{\infty} \frac{(j\eta)^k}{k!} \lambda_k. \tag{5.7}$$

On the other hand, another representation is known for $\ln \Theta(\eta)$ (see, e.g., [10, 12]), namely,

$$\ln \Theta (\eta) = \ln \left[1 + \sum \frac{(j\eta)^k}{k!} \alpha_k\right], \tag{5.8}$$

which is obtained by the power-series expansion of the exponential $\exp(j\eta F)$, followed by statistical averaging. The relations (5.7) and (5.8) form an identity, making it possible to link α_k, μ_k, and λ_k. Thus, in particular,

$$\left.\begin{array}{l} \lambda_1 = \alpha_1 = \langle F \rangle, \\ \lambda_2 = \alpha_2 - \alpha_1^2 = \mu_2 = \sigma_F^2, \\ \lambda_3 = \alpha_3 - 3\alpha_1\alpha_2 + 2\alpha_1^3 = \mu_3, \\ \lambda_4 = \alpha_4 - 3\alpha_2^2 - 4\alpha_1\alpha_2 + 12\alpha_1^2\alpha_2 - 6\alpha_1^4 = \mu_4 - 3\mu_2^2 \end{array}\right\} \tag{5.9}$$

etc.

Consequently, the semi-invariants, and hence the moments about the origin and about the mean for the distribution $W(F)$, may be computed on the basis of the one-dimensional theorem of superposition of stochastic perturbations.

We now consider briefly the proof of the validity of Eq.(5.4), which comprises the main content of the given theorem. We write the conditional characteristic function $\Theta(\eta / 1)$ for any i-th perturbation entering into the sum (5.1):

$$\Theta(\eta/1) = \langle \exp [j \, a_i \eta \, s \, (t - t_i)]\rangle. \tag{5.10}$$

For the sum of n perturbations, in view of their statistical independence, which is one of the conditions for the validity of the law (5.2), we may write

$$\Theta (\eta/n) = \Theta^n (\eta/1).$$

The conditional probability density $W(F/n)$ is found as a Fourier transform of the conditional characteristic function, i.e.,

$$W (F/n) = (1/2\pi) \int_{-\infty}^{\infty} \Theta^n(\eta/1) \exp (-jF\eta) \, d\eta .^* \tag{5.11}$$

Moreover, if we average $W(F/n)$ over all possible n, the distribution $W(F)$ is obtained:

$$W (F) = \sum_{n=0}^{\infty} P (n) W (F/n).$$

---

*The coefficient $\frac{1}{2}\pi$ in front of the integral expression (5.11) could have another value; the only important thing is that the product of the coefficients be equal to $\frac{1}{2}\pi$ in the direct and inverse Fourier transforms connecting the probability density and characteristic function.

Taking (5.11) into account, we arrive at the following relation for W(F):

$$W(F) = (1/2\pi) \int_{-\infty}^{\infty} \sum_{n=0}^{\infty} P(n) \Theta^n (\eta/1) \exp(-jF\eta) \, d\eta. \tag{5.12}$$

On the other hand, by definition,

$$W(F) = (1/2\pi) \int_{-\infty}^{\infty} \Theta(\eta) \exp(-jF\eta) \, d\eta.$$

Comparing this relation with (5.12), we find that the characteristic function $\Theta(\eta)$ of the investigated process (5.1) is determined in the form

$$\Theta(\eta) = \sum_{n=0}^{\infty} P(n) \Theta^n (\eta/1). \tag{5.13}$$

Taking into account that P(n) has the Poisson distribution (5.2), and summing according to (5.13), we fine that

$$\Theta(\eta) = \exp \{\langle n_1 \rangle \ T \ [\Theta(\eta/1) - 1]\}.$$

We go next to $\ln \Theta(\eta)$, whereupon we obtain

$$\ln\Theta(\eta) = \langle n_1 \rangle \ T [\Theta(\eta/1) - 1]. \tag{5.14}$$

We turn now to Eq. (5.10) and write an expression in explicit form for $\Theta(\eta/1)$. For this we need to average the stochastic variables a_i and t_i over all of their possible values. We realize the averaging of a_i by introducing the probability density W(a). For the averaging of t_i we use the following approach. Since the law (5.2) presupposes that the quantity $t - t_i$ is distributed in a sufficiently small interval $(-T/2, T/2)$ uniformly, we can find the time average, assuming that

$$W(t - t_i) = 1/T, \qquad -T/2 \leqslant t - t_i < T/2.$$

Then for $\Theta_T(\eta/1)$ we obtain*

$$\Theta_T(\eta/1) = \int_{-\infty}^{\infty} W(a)(1/T) \int_{-T/2}^{T/2} \exp[j a\eta s(t - t_i)] \, d(t - t_i) \, da,$$

where the parameter t is considered fixed in this case.

We substitute this expression into (5.14). Going over to the new variable of integration and expanding the exponential in the integrand in a power series, we find

$$\ln \Theta_T(\eta) = \sum_{i=1}^{\infty} \frac{(j\eta)^k}{k!} \langle n_1 \rangle \langle a^k \rangle \int_{-T/2}^{T/2} s^k(t) \, dt, \tag{5.15}$$

where $\langle a^k \rangle$ is determined according to (5.3).

---

* The subscript T of the characteristic function $\Theta_T(\eta/1)$ indicates here and elsewhere the requirement that the interval of integration $(-T/2, T/2)$ extend over all possible values of t_i.

If the integral $\int_{-T/2}^{T/2} s^k(t)\,dt$ converges for any T and integers $k \geq 1$, then in the limit as T →

∞, recognizing that

$$\lim_{T \to \infty} \Theta_T(\eta) = \Theta(\eta),$$

we obtain from (5.15)

$$\ln \Theta(\eta) = \sum_{k=1}^{\infty} \frac{(i\eta)^k}{k!} \langle n_1 \rangle \langle a^k \rangle \int_{-\infty}^{\infty} s^k(t)\,dt. \qquad (5.16)$$

Comparing the series (5.7) and (5.16), we arrive at the desired expression (5.4) for the semi-invariants λ_k.

We now explain the meaning of the assumptions that were made in going from (5.15) to (5.16).

When we assumed

$$\lim_{T \to \infty} \int_{-T/2}^{T/2} s^k(t)\,dt = \int_{-\infty}^{\infty} s^k(t)\,dt,$$

we essentially had in mind the following considerations:

The average number of perturbations $\langle n_1 \rangle$ occurring per unit time does not depend on the time in an interval of the order of the effective duration of the perturbations.

The moments of the amplitude distribution $\langle a^k \rangle$ do not depend on the time in the same interval.

The function s(t) decays so rapidly that it becomes integrable for any integers $k \geq 1$.

Consequently, for the proof of the theorem we have not at any time dropped the dependence of $\langle n_1 \rangle$ and $\langle a^k \rangle$ on the time t, having stated only that this dependence is sufficiently weak in the interval $(t - \delta_{ef}/2,\ t + \delta_{ef}/2)$, where δ_{ef} is the effective duration of the perturbation. In this sense, the relation (5.4) may be written in the more general form

$$\lambda_k(t) \approx \langle n_1(t) \rangle \langle a^k(t) \rangle \int_{-\infty}^{\infty} s^k(t)\,dt, \qquad (5.17)$$

which implies the nonstationarity of the process. This type of nonstationarity, when within the limits of the duration of the perturbation, the statistical characteristics of the process vary comparatively slowly with time, leads to the notion of so-called quasi-stationary processes, or processes reducible to stationary processes [19]. Certain features of such processes are discussed later in Sec. 9.

§6. Two-Dimensional Theorem of the Superposition of Stochastic Perturbations and Corollaries

Here we examine two stochastic processes $F_1(t)$ and $F_2(t)$, the states $F_{1n}(t)$ and $F_{2n}(t)$ of which are specified in the form

$$\left.\begin{aligned} F_{1n}(t) &= \sum_{i=1}^{n} a_i s_1(t - t_i), \\ F_{2n}(t) &= \sum_{j=1}^{n} a_j s_2(t - t_j), \end{aligned}\right\} \tag{6.1}$$

where the quantities t_i, t_j, a_i, and a_j are the random onset times and amplitudes of the perturbations; the functions $s_1(t)$ and $s_2(t)$, in general, may differ in form and are separated, for example, by a time shift τ.

The two-dimensional theorem of superposition of perturbations is formulated as follows:

The $(k + l)$-th order semi-invariants λ_{kl} of the two-dimensional distribution $W(F_1, F_2)$ of the investigated processes are defined in the form

$$\lambda_{kl} = \langle n_1 \rangle \langle a^{k+l} \rangle \int_{-\infty}^{\infty} S_1^k(t) S_2^l(t)\, dt,^* \tag{6.2}$$

where, as before, $\langle n_1 \rangle$ is the average number of perturbations occurring per unit time, $\langle a^{k+l} \rangle$ is defined by an equation analogous to (5.3):

$$\langle a^{k+l} \rangle = \int_0^{\infty} a^{k+l} W(a)\, da. \tag{6.3}$$

The semi-invariants λ_{kl} of the two-dimensional distributions are, by definition [10], coefficients of the series expansion of the logarithm of the two-dimensional characteristic function $\Theta(\eta_1, \eta_2)$:

$$\ln \Theta(\eta_1, \eta_2) = \sum_{k=0}^{\infty} \sum_{l=0}^{\infty} \frac{(i\eta_1)^k (i\eta_2)^l}{k!\, l!} \lambda_{kl}. \tag{6.4}$$

The semi-invariants are related to the product moments α_{kl} and μ_{kl} about the origin and about the mean for the two-dimensional distribution $W(F_1, F_2)$. The moments α_{kl} and μ_{kl} are defined by the relations

$$\alpha_{kl} = \int_{-\infty}^{\infty} \int_{-\infty}^{\infty} F_1^k F_2^l W(F_1, F_2)\, dF_1 dF_2, \tag{6.5}$$

$$\mu_{kl} = \int_{-\infty}^{\infty} \int_{-\infty}^{\infty} (F_1 - \langle F_1 \rangle)^k (F_2 - \langle F_2 \rangle)^l W(F_1, F_2)\, dF_1 dF_2. \tag{6.6}$$

---

*In [51] the proof of this theorem is carried out for the case of equal perturbation amplitudes, when $a = 1$. However, the theorem is readily generalized for random amplitudes.

The relationship between λ_{kl}, α_{kl}, and μ_{kl} may be found from the general expression (6.4) and the representation of $\ln \Theta (\eta_1, \eta_2)$ in the form

$$\ln \Theta (\eta_1, \eta_2) = \ln \left[1 + \sum_{k=0}^{\infty} \sum_{l=0}^{\infty} \frac{(j\eta_1)^k (j\eta_2)^l}{k! \, l!} \, \alpha_{kl} \right]. \tag{6.7}$$

This relation is obtained from the definition of the two-dimensional characteristic function

$$\Theta(\eta_1, \eta_2) = \langle \exp (j\eta_1 F_1 + j\eta_2 F_2) \rangle$$

and the power-series expansion of the exponential $\exp (j\eta_1 F_1 + j\eta_2 F_2)$ with subsequent term-by-term statistical averaging.

From a comparison of Eqs. (6.4) and (6.7), taking (6.5) and (6.6) into account, we obtain the following series of relations:

$$\left.\begin{aligned}
\lambda_{01} &= \alpha_{01} = \langle F_1 \rangle, \\
\lambda_{10} &= \alpha_{10} = \langle F_2 \rangle, \\
\lambda_{11} &= \alpha_{11} - \alpha_{01}\alpha_{10} = \mu_{11} = \langle F_1 F_2 \rangle - \langle F_1 \rangle \langle F_2 \rangle, \\
\lambda_{02} &= \alpha_{02} - \alpha_{01}^2 = \mu_{02} = \sigma_{F_1}^2, \\
\lambda_{20} &= \alpha_{20} - \alpha_{10}^2 = \mu_{20} = \sigma_{F_2}^2, \\
\lambda_{12} &= \alpha_{12} - 2\alpha_{01}\alpha_{11} - \alpha_{10}\alpha_{02}, \\
\lambda_{21} &= \alpha_{21} - 2\alpha_{10}\alpha_{11} - \alpha_{01}\alpha_{20}, \\
\lambda_{03} &= \mu_{03} = \langle (F_1 - \langle F_1 \rangle)^3 \rangle, \\
\lambda_{30} &= \mu_{30} = \langle (F_2 - \langle F_2 \rangle)^3 \rangle
\end{aligned}\right\} \tag{6.8}$$

etc.

The proof of the two-dimensional theorem of superposition of stochastic perturbations may be accomplished by the method of characteristic functions, in a manner analogous to that used in the proof for the one-dimensional case in Sec. 5. We give the basic stages of this proof, emphasizing the underlying assumptions.

We form the expression for the joint conditional characteristic function $\Theta (\eta_1, \eta_2 / 1)$ of the pair of perturbations:

$$\Theta(\eta_1, \eta_2/1) = \langle \exp[j\eta_1 a_i s_1(t - t_i) + j\eta_2 a_i s_2(t - t_j)] \rangle. \tag{6.9}$$

The characteristic function $\Theta(\eta_1, \eta_2 / n)$ corresponding to the sums (6.1) of pairwise statistically independent perturbations is equal to

$$\Theta(\eta_1, \eta_2/n) = \Theta^n(\eta_1, \eta_2/1). \tag{6.10}$$

Since, by assumption, the perturbations are pairwise statistically independent, we have for $i \neq j$,

$$\left.\begin{aligned}
\langle a_i a_j s_1 (t - t_i) s_1 (t - t_j) \rangle &= 0, \\
\langle a_i a_j s_2 (t - t_i) s_2 (t - t_j) \rangle &= 0, \\
\langle a_i a_j s_1 (t - t_i) s_2 (t - t_j) \rangle &= 0.
\end{aligned}\right\} \tag{6.11}$$

We note that the set (6.11) reflects the independence of the onset times of the perturbations and for the analysis of reverberation signals implies the absence of statistical correlation between the spatial coordinates of the scatterers.

The conditional two-dimensional probability density $W(F_1, F_2/n)$ with allowance for (6.10) is equal to

$$W(F_1, F_2/n) = (1/2\pi)^2 \int_{-\infty}^{\infty} \int_{-\infty}^{\infty} \exp(-j\eta_1 F_1 - j\eta_2 F_2) \Theta^n(\eta_1, \eta_2/1) \, d\eta_1 d\eta_2. \tag{6.12}$$

We average $W(F_1, F_2/n)$ over all n:

$$W(F_1, F_2) = \sum_{n=0}^{\infty} P(n) W(F_1, F_2/n). \tag{6.13}$$

We also make use of the general relation

$$W(F_1, F_2) = (1/2\pi)^2 \int_{-\infty}^{\infty} \int_{-\infty}^{\infty} \exp(-j\eta_1 F_1 - j\eta_2 F_2) \Theta_T(\eta_1, \eta_2) \, d\eta_1 d\eta_2, \tag{6.14}$$

which relates the two-dimensional probability density to its characteristic function.

Substituting (5.2) and (6.12) into (6.13), we arrive at the expression*

$$W_T(F_1, F_2) = (1/2\pi)^2 \int_{-\infty}^{\infty} \int_{-\infty}^{\infty} \exp(-j\eta_1 F_1 - j\eta_2 F_2) \times$$

$$\times \exp(-\langle n_1 \rangle T) \sum_{n=0}^{\infty} \frac{[\langle n_1 \rangle T \Theta(\eta_1, \eta_2/1)]^n}{n!} \, d\eta_1 d\eta_2,$$

and comparing the latter with (6.14), we obtain for the characteristic function $\Theta(\eta_1, \eta_2)$

$$\Theta_T(\eta_1, \eta_2) = \exp(-\langle n_1 \rangle T) \sum_{n=0}^{\infty} \frac{[\langle n_1 \rangle T \Theta(\eta_1, \eta_2/1)]^n}{n!}.$$

Since the latter expression represents a power-series expansion of an exponential, clearly for $\Theta_T(\eta_1, \eta_2)$ we have

$$\Theta_T(\eta_1, \eta_2) = \exp\{\langle n_1 \rangle T [\Theta(\eta_1, \eta_2/1) - 1]\}.$$

The statistical averaging in (6.9) and the transition from $\Theta_T(\eta_1, \eta_2)$ to $\ln\Theta(\eta_1, \eta_2)$ as $T \to \infty$ is realized as follows:

$$\ln\Theta(\eta_1, \eta_2) = \lim_{T \to \infty} \langle n_1 \rangle T \int_{-\infty}^{\infty} \int_{-\infty}^{\infty} W(a_i, a_i)(1/T^2) \int_{-T/2}^{T/2} \exp[j\eta_1 a_i s_1(t_i) + j\eta_2 a_i s_2(t_i)] \, dt_i da_i da_i. \tag{6.15}$$

---

* The subscript T of the probability density $W_T(F_1, F_2)$ also indicates the requirement for a further limiting transition as $T \to \infty$.

We now take into account the conditions (6.11), expanding the exponential in the integrand of the latter expression in a power series and passing to the limit as $T \to \infty$. This is possible under the condition that the quantities $<n_1>$ and $<a^{k+l}>$ are constant. Then for $\ln \Theta(\eta_1, \eta_2)$ we obtain, on the basis of (6.15),

$$\ln \Theta(\eta_1, \eta_2) = \sum_{k=0}^{\infty} \sum_{l=0}^{\infty} \frac{(j\eta_1)^k}{k!} \frac{(j\eta_2)^l}{l!} \langle n_1 \rangle \langle a^{k+l} \rangle \int_{-\infty}^{\infty} s_1^k(t) s_2^l(t) \, dt. \qquad (6.16)$$

The validity of the given theorem follows from a comparison of (6.16) with the general expansion (6.4), i.e., we arrive at the relation (6.2) for the semi-invariants λ_{kl} of two-dimensional distributions.

We note that if $<a^{k+l}>$ and $<n_1>$ are comparatively slowly varying functions of the time but the assumptions by which the transition to the limit as $T \to \infty$ was tenable are not fulfilled, then (6.2) must be rewritten in the form

$$\lambda_{kl}(t) \approx \langle n_1(t) \rangle \langle a^{k+l}(t) \rangle \int_{-\infty}^{\infty} s_1^k(t) s_2^l(t) \, dt. \qquad (6.17)$$

The relations (5.4), (5.17), (6.2), and (6.7) defining the semi-invariants λ_k and λ_{kl} of the one-dimensional and two-dimensional distributions describe with sufficient completeness the statistical characteristics of processes representing the summation of stochastic perturbations.

It is natural that one should be able to derive the semi-invariants λ_k from the λ_{kl} by letting $l = 0$ (or, equivalently, $k = 0$), whereupon

$$\lambda_k = \lambda_{k0}.$$

It is possible, on the basis of these relations, to calculate such characteristics as the variance of fluctuations, the autocorrelation function, the cross-correlation functions, coefficients of skewness, excess, etc.

We next consider certain corollaries of the theorems of superposition of stochastic perturbations.

<u>Corollary 1.</u> Variance of the process. If $F_1(t) = F_2(t)$, on the basis of the relations (6.2) and (6.8) we have for the variance σ_F^2 of the investigated process

$$\sigma_F^2 = \langle n_1 \rangle \langle a^2 \rangle \int_{-\infty}^{\infty} s^2(t) \, dt. \qquad (6.18)$$

The variance σ_F^2 can be obtained directly and from (5.4) for $k = 2$.

<u>Corollary 2.</u> Autocorrelation function. The autocorrelation function $B_F(\tau)$ can be found if we assume that

$$F_1(t) = F(t),$$
$$F_2(t) = F(t + \tau).$$

Then from (6.2) and (6.8) we obtain for $B_F(\tau)$,

$$B_F(\tau) = \langle n_1 \rangle \langle a^2 \rangle \int_{-\infty}^{\infty} s(t)\, s(t + \tau)\, dt. \qquad (6.19)$$

The correlation coefficient $R_F(\tau)$ will clearly be equal to

$$R_F(\tau) = \frac{\displaystyle\int_{-\infty}^{\infty} s(t)\, s(t + \tau)\, dt}{\displaystyle\int_{-\infty}^{\infty} s^2(t)\, dt}. \qquad (6.20)$$

It is interesting to note that if the parameters $\langle n_1 \rangle$ and $\langle a^2 \rangle$ depend on the time, then

$$B_F(\tau, t) = \langle n_1(t) \rangle \langle a^2(t) \rangle \int_{-\infty}^{\infty} s(t)\, s(t + \tau)\, dt.$$

However, the correlation coefficient $R_F(\tau)$ is not a function of the time in this case, and is defined by the relation (6.20).

Corollary 3.   Joint frequency–time correlation function.   We mean by the joint frequency–time correlation function $b_F(\tau, \Omega)$

$$b_F(\tau, \Omega) = |\langle F^*(t, \Omega)\, F(t + \tau) \rangle|, \qquad (6.21)$$

where $F^*(t, \Omega)$ is the complex conjugate process, the spectrum of which is shifted by an amount Ω relative to the spectrum of the process $F(t)$.

It is essential for the determination of $b_F(\tau, \Omega)$ in proving the theorem of semi-invariants λ_{kl} (for $k = l = 1$) to represent the initial processes (6.1) in the form

$$F^*(t, \Omega) = \sum_{i=0}^{\infty} a_i s(t - t_i)\, \exp [j\Omega(t - t_i)],$$

$$F(t + \tau) = \sum_{j=0}^{\infty} a_j s(t + \tau - t_j).$$

The joint correlation function $b_F(\tau, \Omega)$ is calculated in a manner analogous to that in the derivation of the relation (6.2) for λ_{kl} and turns out to be equal to

$$b_F(\tau, \Omega) = \langle n_1 \rangle \langle a^2 \rangle \left| \int_{-\infty}^{\infty} s^*(t)\, s(t + \tau)\, \exp(j\Omega t)\, dt \right|. \qquad (6.22)$$

Of course, in this case the elementary perturbations $s^*(t - t_i)$ and $s(t + \tau - t_j)$ must also be represented in complex form.   The joint correlation function $b_F(\tau, \Omega)$ may be interpreted, in particular, as a measure of the mutual statistical correlation of reverberation signals in the transmission of pulses having the same form and spectra with different middle frequencies.

Corollary 4.   Correlation of perturbations with random form.   Let the perturbations, as before, have the same form and let the pulses be represented as segments of a stochastic process.

To calculate the correlation function of the process F(t) it is necessary to carry out an additional statistical averaging of the product $s(t)s(t + \tau)$. For $B_F(\tau)$ we have in this case

$$B_F(\tau) = \langle n_1 \rangle \langle a^2 \rangle \int_{-\infty}^{\infty} \langle s(t)\, s(t + \tau) \rangle \, dt. \qquad (6.23)$$

If now we write each perturbation $s(t)$ as a stationary stochastic process $x(t)$ multiplied by some determinant normalized function $s_0(t)$, then

$$\langle s(t)\, s(t + \tau) \rangle = \langle x(t)x(t + \tau) \rangle \, s_0(t)\, s_0(t + \tau). \qquad (6.24)$$

For $B_F(\tau)$ we obtain the following relation on the basis of (6.23) and (6.24):

$$B_F(\tau) = \langle n_1 \rangle \langle a^2 \rangle B_x(\tau) \int_{-\infty}^{\infty} s_0(t)\, s_0(t + \tau)\, dt, \qquad (6.25)$$

where

$$B_x(\tau) = \langle x(t)x(t + \tau) \rangle = \sigma_x^2 R_x(\tau) \qquad (6.26)$$

is the autocorrelation function of the process $x(t)$ and, by virtue of its stationarity, does not depend on the time. We arbitrarily set the variance σ_x^2 equal to unity or denote $<a^2>\sigma_x^2 = <a_x^2>$. Then, taking (6.25) and (6.26) into account, we arrive at the relation

$$B_x(\tau) = \langle n_1 \rangle \langle a_x^2 \rangle R_x(\tau) \int_{-\infty}^{\infty} s_0(t)\, s_0(t + \tau)\, dt. \qquad (6.27) \;.$$

For the correlation coefficient $R_F(\tau)$, clearly, we have

$$R_F(\tau) = R_x(\tau) \frac{\displaystyle\int_{-\infty}^{\infty} s_0(t)\, s_0(t + \tau)\, dt}{\displaystyle\int_{-\infty}^{\infty} s_0^2(t)\, dt} . \qquad (6.28)$$

Corollary 5. Coefficients of skewness and excess. As a characteristic of the form of one-dimensional distributions, in addition to the mean and variance, the coefficients of skewness γ_a and excess γ_e are introduced, which are directly related to the semi-invariants λ_k and central moments about the mean μ_k [10]:

$$\gamma_a = \lambda_3/\lambda_2^{3/2} = \mu_3/\sigma^3, \qquad (6.29)$$

$$\gamma_e = \lambda_4/\lambda_2^2 = \frac{\mu_4}{\sigma^4} - 3. \qquad (6.30)$$

For a process whose odd-order semi-invariants are equal to zero, the coefficient of skewness is also equal to zero. This means that the distribution W(F) will have a zero mean value ($\alpha_1 = 0$) and will be symmetrical with respect to its center, so that W(F) = W(−F).

The coefficient of excess γ_e may be computed on the basis of the general relation (5.4) and the definition (6.30).

§7. Generalization of the Two-Dimensional Theorem of the Superposition of Stochastic Perturbations

So far we have been considering processes whose states are represented as sums of perturbations with random amplitudes and onset times. For the solution of quite a large number of problems, however, it is important to investigate the properties of processes of a more general type, when the perturbations depend, in addition, on some set of stochastic parameters ξ, and the states $F_{1n}(t)$ and $F_{2n}(t)$ are defined in the form (4.5), i.e.,

$$\left.\begin{aligned}
F_{1n}(t) &= \sum_{i=0}^{n} a_i s_1(t - t_i,\ \xi_i), \\
F_{2n}(t) &= \sum_{j=0}^{n} a_j s_2(t - t_j,\ \xi_j).
\end{aligned}\right\} \tag{7.1}$$

Here ξ_i and ξ_j could characterize, for example, the duration of the perturbations, their form, spectral midfrequency, etc.

For the stochastic variables a_i, a_j, t_i, t_j, and n we assume that they obey the same distribution laws in general that were assumed above in Secs. 5 and 6.

Let us find expressions for the semi-invariants λ_{kl} of the joint distribution $W(F_1, F_2)$ of the investigated processes $F_1(t)$ and $F_2(t)$, the states of which are represented in the form (7.1).

We specify the probability density $W(\xi)$, which may be multidimensional in general: $W(\xi_1, \xi_2, \ldots, \xi_k)$.

For the proof of the generalized theorem of semi-invariants it is convenient to apply the method of characteristic functions presented in Sec. 6, the only difference being that in deriving the final expression for the semi-invariants it is necessary to carry out an additional statistical averaging over the set of stochastic parameters ξ.

Following this method, we readily obtain for the semi-invariants λ_{kl}

$$\lambda_{kl} = \langle n_1 \rangle \langle a^{k+1} \rangle \int_{\xi} W(\xi) \int_{-\infty}^{\infty} s_1^k(t, \xi)\, s_2^l(t, \xi)\, dt\, d\xi. \tag{7.2}$$

We note that Eq. (6.2) emerges from the expression (7.2) as a special case. Thus, if ξ is a set of nonstochastic parameters ξ_0, then

$$W(\xi) = \delta(\xi - \xi_0), \qquad \xi_0 = \mathrm{const}$$

and from (7.2) we obtain the equation

$$\lambda_{kl} = \langle n_1 \rangle \langle a^{k+l} \rangle \int_{-\infty}^{\infty} s_1^k(t, \xi_0)\, s_2^l(t, \xi_0)\, dt,$$

which is like (6.2).

For k = 1 and l = 1, Eq. (7.2) leads to an expression for the correlation function $B_{12}(z)$:

$$B_{12}(z) = \langle n_1 \rangle \langle a^2 \rangle \int_{\xi} W(\xi) \int_{-\infty}^{\infty} s_1(t_1, \xi)\, s_2(t, \xi)\, dt\, d\xi, \tag{7.3}$$

where z is the set of parameters characterizing the relationship between the processes $F_1(t)$ and $F_2(t)$.

The relations obtained above enable one to investigate the influence on the statistical characteristics of reverberation of such factors as the motion of the scatterers and the acoustic arrays, as well as to investigate the space—time correlation of reverberation.

§ 8. Representation of the Probability Density in Series Form

We will have occasion later on to encounter the following problem in connection with the determination of the one-dimensional distributions of reverberation signals: to find an expression for the probability density when the parameters of the distribution (moments or semi-invariants) are known.

A suitable approach to the solution of this problem is to expand the unknown probability density in a series whose coefficients will be related directly to the above-mentioned parameters of the distribution.

Such a representation of the probability density is useful for analyzing the statistical characteristics of reverberation signals, in that it is possible, on the basis of the theorems of superposition of stochastic perturbations, to calculate with relative ease a certain number of the semi-invariants, which determine the terms of the series and make it possible to find the distribution law with fair accuracy.

The probability density is represented in the following form suitable for our purposes:

$$W(F) = c_0 \varphi(F) + \frac{c_1}{1!} \varphi'(F) + \frac{c_2}{2!} \varphi''(F) + \dots, \qquad (8.1)$$

where c_0, c_1, c_2, ... are constant coefficients, $\varphi(F)$ is some function forming the kernel of the given expansion.

For the majority of problems the most suitable and widely used type of kernel $\varphi(F)$ of the expansion is a squared exponential. This allows direct comparison of the probability density $W(F)$ with a normal distribution and is particularly convenient if the investigated stochastic variable is formed as the result of superposition of independent of very slightly dependent components.

Let X by a normalized stochastic variable

$$X = (F - \langle F \rangle)/\sigma_F, \qquad (8.2)$$

where $<F>$ and σ_F are the mean and mean-square values, respectively, of the initial stochastic function F(t) defining some physical process. In this case, the kernel of the expansion is best given in the form

$$\varphi(X) = \left(1/\sqrt{2\pi}\right) \exp\left(-X^2/2\right). \qquad (8.3)$$

Now the probability density W(F) represented by the series (8.1) goes over to the dimensionless probability density $W_X(X)$.

Considering the definition (8.3), it is easily shown that the function $\varphi(X)$ is related to its k-th derivative by the expression

$$\varphi^{(k)}(X) = (-1)^k H_k(X) \varphi(X), \qquad (8.4)$$

where $H_k(X)$ is a k-th order Hermite polynomial. Differentiating $\varphi(X)$ directly, or making use of the familiar recursion relation

$$H_{k+1}(X) = XH_k(X) - kH_{k-1}(X),$$

we obtain for the first six Hermite polynomials

$$\left.\begin{aligned}
H_0(X) &= 1, \\
H_1(X) &= X, \\
H_2(X) &= X^2 - 1, \\
H_3(X) &= X^3 - 3X, \\
H_4(X) &= X^4 - 6X^2 + 3, \\
H_5(X) &= X^5 - 10X^3 + 15X, \\
H_6(X) &= X^6 - 15X^4 + 45X^2 - 15.
\end{aligned}\right\} \tag{8.5}$$

The coefficients $c_0, c_1, c_2, \ldots$ of the series (8.1) are determined as follows. It is required to substitute into (8.1), in place of the derivatives $\varphi^{(k)}(X)$, their expressions in terms of the Hermite polynomials according to (8.4), then to multiply the right- and left-hand sides of the series by $H_l(X)$ and integrate them over all possible values of X. Then, invoking the orthogonality condition

$$\left(1/\sqrt{k!l!}\right) \int_{-\infty}^{\infty} H_k(X) H_l(X) \varphi(X)\, dX = \begin{cases} 1 & \text{for } k = l, \\ 0 & \text{for } k \neq l, \end{cases} \tag{8.6}$$

we find for the coefficients c_k

$$c_k = (-1)^k \int_{-\infty}^{\infty} H_k(X) W_x(X)\, dX. \tag{8.7}$$

The Hermite polynomials $H_k(X)$ are k-th degree polynomials, which means, by virtue of (8.7) that the coefficients c_k must be linear functions of the moments α_i or μ_i ($i \leq k$) of the sought-after probability density $W_x(X)$. Consequently, bearing Eqs. (8.2) and (8.7) in mind, plus the fact that

$$W_x(X) = \sigma_F W(\langle F \rangle + X\sigma_F),$$

$$\langle X^k \rangle = \mu_k/\sigma_F^k = \int_{-\infty}^{\infty} X^k W_x(X)\, dX,$$

we obtain for the c_k, through k = 6,

$$\left.\begin{aligned}
c_0 &= 1, \\
c_1 &= 0, \\
c_2 &= 0, \\
c_3 &= -\mu_3 \sigma_F^{-3}, \\
c_4 &= \mu_4 \sigma_F^{-4} - 3, \\
c_5 &= -(\mu_5 \sigma_F^{-5} - 10\mu_3 \sigma_F^{-3}), \\
c_6 &= \mu_6 \sigma_F^{-6} - 15\mu_4 \sigma_F^{-4} + 30.
\end{aligned}\right\} \tag{8.8}$$

Hence, with (8.8) in mind, the series (8.1) may be represented in the following form for the density $W_x(X)$:

$$W_x(X) = \varphi(X) - \frac{\mu_3 \varphi^{(3)}(X)}{3! \sigma_F^3} + \frac{1}{4!} \left(\frac{\mu_4}{\sigma_F^4} - 3 \right) \varphi^{(4)}(X) -$$
$$- \frac{1}{5!} \left(\frac{\mu_5}{\sigma_F^5} - 10 \frac{\mu_3}{\sigma_F^3} \right) \varphi^{(5)}(X) + \frac{1}{6!} \left(\frac{\mu_6}{\sigma_F^6} - 15 \frac{\mu_4}{\sigma_F^4} + 30 \right) \varphi^{(6)}(X) - \cdots \qquad (8.9)$$

Replacing the derivatives $\varphi^{(k)}(X)$ in accordance with (8.4), we have

$$W_x(X) = \left(1/\sqrt{2\pi} \right) \exp\left(-X^2/2 \right) \left[1 + \frac{1}{3!} \frac{\mu_3}{\sigma_F^3} H_3(X) + \right.$$
$$+ \frac{1}{4!} \left(\frac{\mu_4}{\sigma_F^4} - 3 \right) H_4(X) + \frac{1}{5!} \left(\frac{\mu_5}{\sigma_F^5} - 10 \frac{\mu_3}{\sigma_F^3} \right) H_5(X) +$$
$$\left. + \frac{1}{6!} \left(\frac{\mu_6}{\sigma_F^6} - 15 \frac{\mu_4}{\sigma_F^4} + 30 \right) H_6(X) + \cdots \right]. \qquad (8.10)$$

It is also possible, making use of (8.2) and (8.10), to write down a series for the probability density $W(F)$ of the quantity F corresponding to the physical process and having a definite dimensionality (e.g., volts, bars, etc.):

$$W(F) = \left(1/\sqrt{2\pi}\sigma_F \right) \exp\left[-(F - \langle F \rangle)^2 / 2\sigma_F^2 \right] \left[1 + \frac{1}{3} \frac{\mu_3}{\sigma_F^3} H_3 \left(\frac{F - \langle F \rangle}{\sigma_F} \right) + \right.$$
$$+ \frac{1}{4!} \left(\frac{\mu_4}{\sigma_F^4} - 3 \right) H_4 \left(\frac{F - \langle F \rangle}{\sigma_F} \right) + \frac{1}{5!} \left(\frac{\mu_5}{\sigma_F^5} - 10 \frac{\mu_3}{\sigma_F^3} \right) H_5 \left(\frac{F - \langle F \rangle}{\sigma_F} \right) +$$
$$\left. + \frac{1}{6!} \left(\frac{\mu_6}{\sigma_F^6} - 15 \frac{\mu_4}{\sigma_F^4} + 30 \right) H_6 \left(\frac{F - \langle F \rangle}{\sigma_F} \right) + \cdots \right] \qquad (8.11)$$

Analogously, proceeding from (8.9), we find the probability density $W(F)$ expressed in terms of the derivatives $\varphi^{(k)}[(F - \langle F \rangle)/\sigma_F]$.

We note that the central moments about the mean, which determine the coefficients in the series given above, may be expressed according to (5.9) in terms of the semi-invariants found on the basis of the theorems of superposition of stochastic perturbations, proceeding from the physical characteristics of the processes investigated.

The speed of convergence of the investigated series may be estimated from the known values of the moments of the distribution μ_k of the semi-invariants λ_k, relying on their behavior as a function of the order k, in particular as $k \to \infty$. Knowledge of these coefficients of the terms of the series makes it possible in principle to find the required number of terms in order for the distribution to be approximated with a prescribed accuracy. The speed of convergence will depend on the "degree of smoothness" of the distribution sought after and the character of its behavior at infinity.

If we retain only the first few terms of the expansions and make use of the coefficients of skewness and excess defined in (6.29) and (6.30), then, for example, the following approximate relations may be written for the series (8.9) and (8.11), grouping their terms in appropriate fashion:

$$W_x(X) \approx \varphi(X) - \frac{1}{3!} \gamma_a \varphi^{(3)}(X) + \frac{1}{4!} \gamma_e \varphi^{(4)}(X) + \frac{10}{6!} \gamma_a^2 \varphi^{(6)}(X), \qquad (8.12)$$

$$W(F) \approx (1/\sqrt{2\pi}\sigma_F) \exp\left[\frac{-(F-\langle F\rangle)}{2\sigma_F^2}\right]\left[1 + \frac{1}{3!}\gamma_a H_3\left(\frac{F-\langle F\rangle}{\sigma_F}\right) + \right.$$
$$\left. + \frac{1}{4!}\gamma_e H_4\left(\frac{F-\langle F\rangle}{\sigma_F}\right) + \frac{10}{6!}\gamma_a^2 H_6\left(\frac{F-\langle F\rangle}{\sigma_F}\right)\right]. \tag{8.13}$$

In many cases of practical interest these series offer sufficient accuracy in calculating the distributions according to the known coefficients of skewness and excess.

The expansions obtained above are known as Edgeworth and Gram-Charlier series and are used (see, e.g., [10, 12, 19]) for the approximate representation of distribution laws, proof of the convergence of distributions to a normal law, etc. Both series are based on a normal distribution law, their only difference being in the grouping of the terms and the modes of calculating the coefficients.

§ 9. Stochastic Processes Reducible to Stationary

We first give certain definitions that are normally used in the analysis of stochastic processes of various types.

A stochastic process V(t) is called stationary in the strict sense* when its statistical properties do not vary when the set of reference times are shifted by an equal amount t'. In particular, if an n-dimensional probability density is given:

$$W(V_1, t_1;\ V_2, t_2;\ \ldots;\ V_n, t_n),$$

the following equation holds for a process stationary in the strict sense:

$$W(V_1, t_1;\ V_2, t_2;\ \ldots;\ V_n, t_n) = W(V_1, t_1+t';\ V_2, t_2+t';\ \ldots;\ V_n, t_n+t). \tag{9.1}$$

Also used is the notion of stationarity in the broad sense, referring to processes for which the correlation function does not depend on the time-reference origin, i.e.,

$$B_V(\tau) = \langle V(t_1)V(t_2)\rangle = \langle V(t_1+t')V(t_2+t')\rangle = \langle V(t)V(t+\tau)\rangle, \tag{9.2}$$

where $\tau = t_2 - t_1$ [we have assumed for simplicity that $<V(t)> = 0$].

In some applications one often encounters stochastic processes that may be expressed fairly simply in terms of stationary processes, at least in the broad sense. Such processes are sometimes called reducible to stationary or quasi-stationary.

The theory of these processes has been developed in [19]; we will consider here only some of their characteristics required in studying the statistical properties of reverberation.

If a stochastic process F(t) is representable in the form

$$F(t) = g(t)V(t), \tag{9.3}$$

where V(t) is a process stationary in the broad sense, g(t) is some determinant function, then the process may be reduced to stationary.

For $<V(t)> = 0$, the mean $<F(t)>$, variance σ_F^2, and correlation function $B_F(t, t+\tau)$ of the process (9.3) are defined as follows:

---

*Sometimes referred to also as stationarity in the narrow sense.

$$\langle F(t) \rangle = g(t) \langle V(t) \rangle = 0, \tag{9.4}$$

$$\sigma_F^2 = g^2(t) \sigma_V^2, \tag{9.5}$$

$$B_F(t, t + \tau) = g(t) g(t + \tau) B_F(\tau), \tag{9.6}$$

where σ_V^2 and $B_V(\tau)$ are the variance and correlation function of the stationary process V(t), respectively.

On the basis of (9.5) and (9.6), we obtain for the correlation coefficient $R_F(\tau)$, defined as

$$R_F(\tau) = \frac{B_F(t, t + \tau)}{\sigma_F(t) \sigma_F(t + \tau)}, \tag{9.7}$$

the following expression:

$$R_F(\tau) = R_V(\tau). \tag{9.8}$$

Equation (9.8) thus permits us to formulate the following condition. If the correlation coefficient $R_F(\tau)$ of the process F(t) depends only on the correlation interval τ, then the process is reducible to stationary.

We now consider the problem of the type of averaging that will permit the above-indicated statistical characteristics of the process F(t) to be ascertained.

On the one hand, the correlation function $B_F(t, t + \tau)$ and variance $\sigma_F^2(t)$ may be obtained by taking the average over the ensemble of states, for example, according to the two-dimensional probability density $W(F, t; F_\tau, t + \tau)$ for t = const.

On the other hand, it is very desirable to use time averaging, since often an adequate number of states of the process F(t) are not at one's disposal. In this case it is instructive to investigate the characteristics of the processes g(t) and V(t) separately, where at least two approaches to such an investigation are possible, namely, averaging of F(t) over a finite interval, and separation of the stationary part of the process (9.3) and investigation of the properties of the process V(t).

In the first instance we have for $\sigma_F^2(t, T)$ and $B_F(t, t + \tau, T)$:

$$\sigma_F^2(t, T) = (1/T) \int_{-T/2}^{T/2} g^2(t) V^2(t) \, dt, \tag{9.9}$$

$$B_F(t, t + \tau, T) = (1/T) \int_{-T/2}^{T/2} g(t) g(\tau + t) V(t) V(t + \tau) \, dt. \tag{9.10}$$

If we now assume that the function g(t) varies imperceptibly on the averaging interval $(-T/2, T/2)$, so that

$$g(t) \approx g(t + \tau), \tag{9.11}$$

then, in particular, the following may be written for $R_F(\tau)$:

$$R_F(\tau) \approx \frac{\int_{-T/2}^{T/2} V(t) V(t+\tau)\,dt}{\int_{-T/2}^{T/2} V^2(t)\,dt}.\qquad(9.12)$$

The interval $(-T/2, T/2)$ must be small enough to ensure the required accuracy of estimation, i.e., small relative fluctuations of the integrals

$$\left(1/T\right) \int_{-T/2}^{T/2} V^2(t)\,dt,\ \ (1/T) \int_{-T/2}^{T/2} V(t) V(t+\tau)\,dt.$$

In the second instance, when the object is to separate out the stochastic process V(t) comprising the stationary part of F(t), it is necessary to perform the following operation, similar to normalization:

$$V(t) = F(t)/g(t),\qquad(9.13)$$

for which the determinate component g(t) must be separated from F(t). This operation can be affected, for example, by means of a partial separation of the functions V(t) and g(t) after a suitable nonlinear transformation of the process F(t) using a feedback system, into the circuit of which the function g(t) is injected; other techniques may also be used.

The indicated algorithm (9.13) is equivalent to the operation of _stationarization_ of the investigated process, and the process V(t) so generated may be called the _stationarized_ process.

Both of the processing modes noted above require the following condition to be met. The spectra or correlation characteristics of the processes V(t) and g(t) must differ appreciably from one another. This happens when the process g(t) varies much more slowly than the stationary part.

There is, of course, one other possibility for the analysis of the given processes, namely combined averaging over the time on the interval $(-T/2, T/2)$ and over the ensemble of N states. Then the correlation coefficient $R_V(\tau)$ may be found by means of the relation

$$R_V(\tau) \approx \frac{\sum_{i=1}^{N} (1/T) \int_{-T/2}^{T/2} F^{(i)}(t) F^{(i)}(t+\tau)\,dt}{\sum_{i=1}^{N} (1/T) \int_{-T/2}^{T/2} [F^{(i)}(t)]^2\,dt},\qquad(9.14)$$

where $i = 1, 2, \ldots, N$ corresponds to the number of states of the stochastic process F(t).

Sea reverberation signals may be classed among processes of the above type in many cases of practical interest, and they may be reduced to stationary in principle, provided certain conditions are met. The role of g(t) in this case is played by the function defining the decay law of the mean reverberation level with time, and the stationary part V(t) is represented by the sum, obtained after normalization, of the elementary scattered signals arriving simultaneously at the point of reception.

It is important to bear in mind one feature of reverberation as a process reducible to stationary. This is the fact that the time variation of the average number of elementary scattered signals, given relatively few of the latter, not only affects the variance of reverberation, but also its one- and two-dimensional distributions, even though the correlation coefficient turns out to be independent of the time. This lays further emphasis on the fact that we are very likely concerned here with the condition of stationarity only in the broad sense, because, for a stationarized process the moments μ_{kl} of the distribution or the semi-invariants λ_{kl} of orders higher than the correlation values may turn out to be dependent on the time.

CHAPTER II

TRANSMITTED SIGNALS AND THEIR CHARACTERISTICS

§ 10. Types of Signals

The transmitted signals may be regarded as some function s(t) having the following form in the general case:

$$s(t) = s_0(t) \cos [\omega_0(t) + \Phi(t) + \varphi_0], \qquad (10.1)$$

where $s_0(t)$ is the envelope, ω_0 is the carrier or central frequency of the signal spectrum, $\Phi(t)$ is a function defining the phase modulation, φ_0 is the initial phase.

From here on we will say that $\varphi_0 = 0$, as this is merely a matter of choice of the reference point of s(t) in time.

If the frequency modulation law $\omega(t)$ is given, $\Phi(t)$ and $\omega(t)$ are interrelated by the well-known relations

$$\left. \begin{array}{l} \Phi(t) = \displaystyle\int_0^t \omega(t')\, dt', \\[2mm] \omega(t) = d\Phi(t)/dt. \end{array} \right\} \qquad (10.2)$$

In some cases it is convenient to write S(t) in complex form, i.e.,

$$s(t) = \operatorname{Re} s_{01}(t) \exp(j\omega_0 t), \qquad (10.3)$$

where $s_{01}(t)$ indicates a function characterizing the complex law of modulation of the signal and including the envelope $s_0(t)$, as well as $\exp[j\Phi(t)]$, which determines the law of phase modulation, i.e.,

$$s_{01}(t) = s_0(t) \exp[j\Phi(t)]. \qquad (10.4)$$

In the investigation of the statistical properties of reverberation we will be concerned primarily with pulse-type signals, i.e., signals whose envelopes are described by functions $s_0(t)$ satisfying the conditions

$$\left. \begin{array}{l} \displaystyle\int_{-\infty}^{\infty} |s_0(t)|\, dt = \mathrm{const} < +\infty, \\[4mm] \displaystyle\int_{-\infty}^{\infty} s_0^2(t)\, dt = \mathrm{const} < +\infty. \end{array} \right\} \qquad (10.5)$$

It sometimes proves adequate to name only the second condition, i.e., finiteness of the signal energy, and to assume that the function $s_0(t)$ does not have any points at which it goes to infinity. Then the requirement that $s_0(t)$ be an integrable function clearly drops out.

We will examine three fundamental types of signals, the transmission of which leads to significant differences in the correlation and spectral properties of the reverberation, namely:

Signals with amplitude modulation;

Signals with frequency modulation;

Noisy signals.

Signals with amplitude modulation are described by a function s(t) of the type

$$s(t) = s_0(t) \cos \omega_0 t. \tag{10.6}$$

Included among signals of this type are pulses with various envelope profiles and pulses modulated in amplitude according to a periodic law; the carrier frequency ω_0 in this case is constant.

The envelopes of some typical signals of the indicated type, normalized to unity, are written as follows:

Rectangular pulse:

$$s_0(t) = 1, \ |t| \leqslant \delta/2; \tag{10.7}$$

Bell-shaped (Gaussian) pulse:

$$s_0(t) = \exp\left[-(t/t_0)^2\right]; \tag{10.8}$$

Exponential pulse:

$$s_0(t) = \exp\left(-t/t_0\right), \ t \geqslant 0. \tag{10.9}$$

Signals with amplitude modulation according to a harmonic law may be described in the form

$$s(t) = \frac{1 + m \cos \Omega_1 t}{1 + m} s_0(t) \cos \omega_0 t, \tag{10.10}$$

where m is the modulation index, which varies over the interval (0, 1), Ω_1 is the modulation frequency. The factor $1/(1 + m)$ is included in (10.10) for normalization of the envelope of the given pulse; the function $s_0(t)$ can in principle assume any form and must satisfy the conditions (10.5).

We next consider signals with frequency modulation. We will subsequently be interested primarily in a linear modulation law $\omega(t)$, namely,

$$\omega(t) = \Delta\omega_M t/t_M; \tag{10.11}$$

the parameters $\Delta\omega_M$ and t_M characterize the frequency deviation and its rate of change. We write for the law of phase variation $\Phi(t)$, taking Eqs. (10.2) and (10.10) into account,

$$\Phi(t) = \frac{\Delta\omega_M t^2}{2t_M}. \tag{10.12}$$

In this case the signal s(t) is defined on the basis of (10.1) and (10.12) as

$$s(t) = s_0(t)\cos(\omega_0 t + \Delta\omega_M t^2/2t_M), \tag{10.13}$$

where the envelope s(t), generally speaking, can be described by any of the relations (10.7)–(10.10).

We next consider what we call noisy signals. A noisy signal s(t) may be written in the form

$$s(t) = s_0(t)\,x(t), \tag{10.14}$$

where $s_0(t)$, as before, is a determinant function satisfying the conditions (10.5), x(t) is a stationary stochastic process. If the latter is a narrow-band process, s(t) may be rewritten in a form analogous to (10.1), i.e.,

$$s(t) = s_0(t)\,E(t)\cos[\omega_0 t + \psi(t)], \tag{10.15}$$

where E(t) and $\psi(t)$ are the envelope and phase of the stochastic process x(t), ω_0 is the central frequency of its spectrum.

The function $s_0(t)$ in Eqs. (10.14) and (10.15) defines the pulse character of noisy signals. If it has the form of a rectangular pulse (10.7), it then follows from (10.15) that

$$s(t) = E(t)\cos[\omega_0 t + \psi(t)], \quad |t| \leqslant \delta/2. \tag{10.16}$$

We may go from this to the following representation of the signal:

$$s(t) = x_c(t)\cos\omega_0 t - x_s(t)\sin\omega_0 t, \quad |t| \leqslant \delta/2, \tag{10.17}$$

where

$$\left. \begin{array}{l} x_c(t) = E(t)\cos\psi(t), \\ x_s(t) = E(t)\sin\psi(t) \end{array} \right\} \tag{10.18}$$

are called the quadrature components of the stochastic process.

Another type of signal is a pulse train, which represents a sequence of several identical pulses $s_1(t)$ one after the other, separated by certain intervals of time. For a signal s(t) of this type, we have the relation

$$s(t) = \sum_{k=0}^{N-1} s_1[t - kT_0 + (N-1)T_0/2], \tag{10.19}$$

where N is the number of pulses in the train, T_0 is their period of repetition.

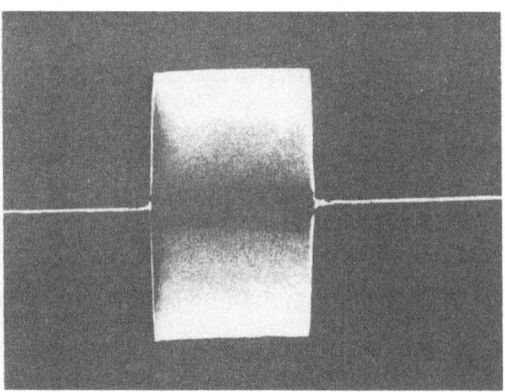

Fig. 10.   Pulse with rectangular envelope.

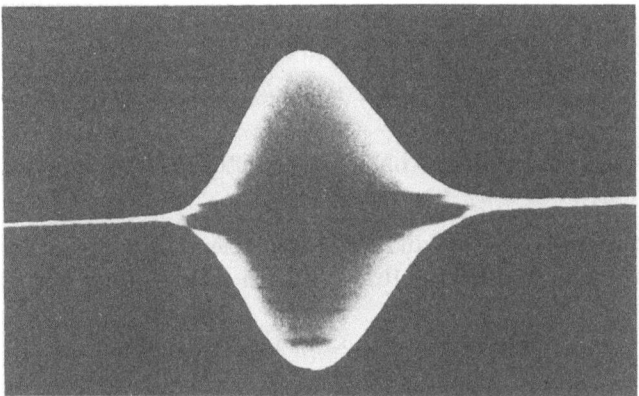

Fig. 11.   Pulse with bell-shaped envelope.

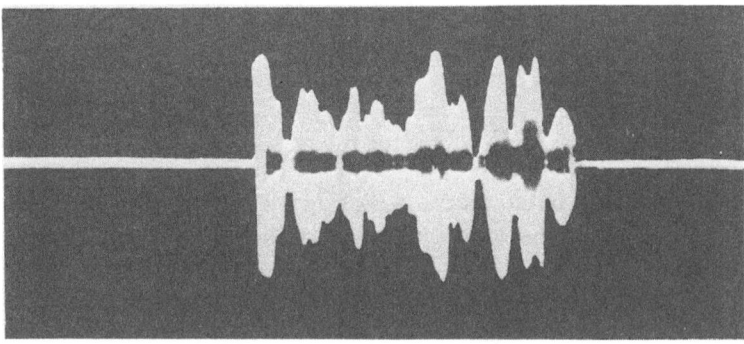

Fig. 12.   Noisy pulse.

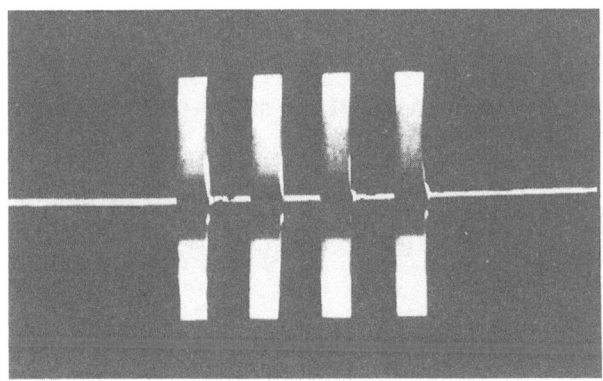

Fig. 13.   Train of rectangular pulses.

In the general case the function $s_1(t)$ may represent any of the functions s(t) discussed above for single pulses having a sinusoidal carrier, frequency-modulated pulses, or noisy pulses.

Oscillograms of some of the given types of signals are shown in Figs. 10-13.

§ 11.   Effective Signal Duration and Width of the Spectrum

In comparing the characteristics of reverberation for the investigation of pulses of various forms, it is desirable to adopt some criterion of their energy equivalence that would lead to an identical mean intensity on the part of the reverberation signals, all other conditions being equal. A suitable criterion would be equality of the signal energies. If the envelope $s_0(t)$ in this case is normalized to unity, the criterion of energy equivalence may be written in the form

$$\delta_{ef} = \int_{-\infty}^{\infty} s_0^2(t)\, dt = \text{const},\tag{11.1}$$

where δ_{ef} is the effective signal duration, which we interpret as the duration of a rectangular pulse of unit amplitude having an energy equal to the energy of the signal with arbitrary envelope $s_0(t)$. Thus, in particular, if $s_0(t)$ corresponds to the rectangular pulse (10.7), then $\delta_{ef} = \delta$.

Equation (11.1) is used to find the effective duration of signals of any form, including noisy signals.

An important characteristic of the signals is their frequency spectrum, which is determined for the determinant function s(t) by the Fourier transform of that function; in other words,

$$g(\omega) = \int_{-\infty}^{\infty} s(t) \exp(-j\omega t)\, dt,\tag{11.2}$$

where in general the spectrum $g(\omega)$ may be complex. The inverse transform of (11.2) holds:

$$s(t) = (1/2\pi) \int\limits_{-\infty}^{\infty} g(\omega) \exp(j\omega t) \, d\omega. \tag{11.3}$$

The relationship between the intensity of the signal envelope and the energy spectrum is set forth in the so-called Parseval relation:

$$\int\limits_{-\infty}^{\infty} s_0^2(t) \, dt = (1/\pi) \int\limits_{0}^{\infty} |g_0(\omega)|^2 d\omega, \tag{11.4}$$

where $g_0(\omega)$ is the spectrum of the signal envelope.

It follows from (11.1) and (11.4) that the effective signal duration can also be found by means of the spectrum

$$\delta_{ef} = (1/\pi) \int\limits_{0}^{\infty} |g_0(\omega)|^2 d\omega. \tag{11.5}$$

In addition to the signal duration, we will be concerned with the effective width of their spectrum Δf_{ef}, by which we mean the following characteristic:

$$\Delta f_{ef} = \frac{\int\limits_{0}^{\infty} |g(\omega)|^2 d\omega}{2\pi |g(\omega)|_{max}^2}, \qquad \Delta \omega_{ef} = 2\pi \Delta f_{ef}, \tag{11.6}$$

where $g(\omega)$ is determined by the relation (11.2), and $|g(\omega)|_{max}$ is the maximum value of the amplitude spectrum of the signal.

Equation (11.6) is applicable in the case when the function s(t) is described by Eq. (10.1), and the effective width of the signal spectrum is determined both by the spectrum $g_0(\omega)$ of the envelope and by the spectrum of the function $\cos[\omega_0 t + \Phi(t)]$ characterizing the frequency modulation of the signal.

In the absence of frequency modulation, Eq. (11.6) must be rewritten in the form

$$\Delta f_{ef} = \frac{\int\limits_{0}^{\infty} |g_0(\omega)|^2 d\omega}{\pi |g_0(\omega)|_{max}^2}. \tag{11.7}$$

Proceeding from (11.4), (11.5), and (11.7), we arrive at the following expression relating Δf_{ef} and δ_{ef}:

$$\Delta f_{ef} \delta_{ef} = \left(\frac{\int\limits_{-\infty}^{\infty} s_0^2(t) \, dt}{\int\limits_{-\infty}^{\infty} s_0(t) \, dt} \right)^2. \tag{11.8}$$

Table 11.1 gives the analytical spectra of the envelopes of pulses with a sinusoidal carrier, the values of their effective duration, and their spectral widths.

Table 11.1. Characteristics of Pulses with Various Envelope Profiles

No.	Type of pulse		Pulse spectrum	δ_{ef}	Δf_{ef}
1	Rectangular	$S_0(t) = 1, \ |t| \leqslant \delta/2$	$\delta \dfrac{\sin(\omega\delta/2)}{\omega\delta/2}$	δ	$1/\delta$
2	Bell-shaped	$S_0(t) = \exp(-t^2/t_0^2)$ $S_0(t) = \exp(-t^2/t_0^2)$ $|t| \leqslant t_1/2$	$\sqrt{\pi}\,t_0 \exp(-\omega^2 t_0^2/4)$ $(\sqrt{\pi}\,t_0/2)[\exp(-\omega^2 t_0^2/4)] \times$ $\times \left\{ \Phi\left[\dfrac{t_1 + j\omega t_0^2}{2t_0}\right] - \right.$ $\left. - \Phi\left[\dfrac{j\omega t_0^2 - t_1}{2t_0}\right] \right\}$	$\sqrt{\dfrac{\pi}{2}}\,t_0$ $\sqrt{\dfrac{\pi}{2}}\,t_0 \Phi\left[\dfrac{t_0}{\sqrt{2}\,t_0}\right]$	$1/\sqrt{2\pi}\,t_0$ $\dfrac{\Phi\left(\dfrac{t_1}{\sqrt{2}\,t_0}\right)}{\sqrt{2\pi}\left[\Phi\left(\dfrac{t_1}{2t_0}\right)\right]^2 t_0}$
3	Cosine to the p-th power	$S_0(t) = \cos^p(\pi t/t_1)$ $|t| \leqslant t_1/2$	$(2t_1/\pi)\dfrac{\cos(\omega t_1/2)}{1 - \left(\dfrac{\omega t_1}{\pi}\right)^2}, \ p = 1$ $\dfrac{4\pi^2}{\omega(4\pi^2 - \omega^2 t_1^2)}\sin(\omega t_1/2),$ $p = 2$	$t_1/2$ $3t_1/8$	$\pi^2/8t_1$ $3/2t_1$
4	Exponential	$S_0(t) = \exp(-t/t_0),$ $t \geqslant 0$ $S_0(t) = \exp(-t/t_0),$ $t \geqslant 0, \ t < t_1$	$t_0/(1 + j\omega t_0)$ $t_0/(1 + j\omega t_0)\{1 - $ $- \exp[-(1 + j\omega t_0)t_1/t_0]\}$	$t_0/2$ $t_0/2[1 - $ $- \exp(-2t_1/t_0)]$	$1/2t_0$ $(^1\!/_2 t_0)\dfrac{1 + \exp(-t_1/t_0)}{1 - \exp(-t_1/t_0)}$
5		$S_0(t) = \dfrac{1 + m\cos\Omega_1 t}{1 + m}$ $|t| \leqslant \delta/2$ $\Omega_1\delta \gg 1$	$\dfrac{\delta}{1+m}\dfrac{\sin(\omega\delta/2)}{\omega\delta/2} +$ $+ \dfrac{m\delta}{2(1+m)}\dfrac{\sin[(\Omega_1 - \omega)\delta/2]}{(\Omega_1 - \omega)\delta/2} +$ $+ \dfrac{m\delta}{2(1+m)}\dfrac{\sin[(\Omega_1 + \omega)\delta/2]}{(\Omega_1 + \omega)\delta/2}$	$\dfrac{\delta}{(1+m)^2}(1 + m^2/2)$	—

The following definition of the probability integral $\Phi(X)$ [9] is used in the table:

$$\Phi(X) = (2/\sqrt{\pi})\int_0^X \exp(-x^2)\,dx. \tag{11.9}$$

Let us look further into the evaluation of the parameters for noisy signals.

The effective duration in this case is defined analogously to (11.1), where $s_0(t)$ is a determinant function governing the pulsed character of the signal.

The effective width of the spectrum Δf_{ef} is expressed in the form

$$\Delta f_{ef} = \frac{\int\limits_{0}^{\infty}\int\limits_{-\infty}^{\infty} G_x(\nu)\,|\,g_0(\omega-\nu)\,|^2\,d\nu\,d\omega}{2\pi\left[\int\limits_{-\infty}^{\infty} G_x(\nu)\,|\,g_0(\omega-\nu)^2\,|\,d\nu\right]_{max}}, \tag{11.10}$$

where $G_x(\omega)$ is the statistical spectrum of the stationary process $x(t)$ governing $s(t)$ according to (10.14), $g_0(\omega)$ is again the spectrum of the function $s_0(t)$. Equation (11.10) leads to the relation (11.7) if the spectrum of the process $x(t)$ is given in the form of a delta function:

$$G_x(\omega) = \sigma_x^2 \delta(\omega - \omega_0),$$

where σ_x^2 is the variance of the process. Hence, considering that

$$\sigma_x^2 \int\limits_{-\infty}^{\infty} |\,g_0(\omega-\nu)\,|^2 \delta(\nu-\omega_0)\,d\nu = \sigma_x^2\,|\,g_0(\omega-\omega_0)\,|^2,$$

we obtain from (11.10)

$$\Delta f_{ef} = \frac{\int\limits_{0}^{\infty} |\,g_0(\omega-\omega_0)\,|^2\,d\omega}{2\pi\,|\,g_0(\omega-\omega_0)\,|_{max}^2} \tag{11.11}$$

Bearing in mind the obvious equality

$$|\,g_0(\omega-\omega_0)\,|_{max}^2 = |\,g_0(\omega)\,|_{max}^2,$$

as well as the fact that for narrow-band processes

$$\int\limits_{0}^{\infty} |\,g_0(\omega-\omega_0)\,|^2\,d\omega \approx 2\int\limits_{0}^{\infty} |\,g_0(\omega)\,|^2\,d\omega,$$

we arrive from (11.11) at the relation (11.7).

The concept of narrow-band signals was used above. Narrow bandwidth, or as it is sometimes called, quasi-harmonicity, is evaluated in terms of the inequality

$$\Delta f_{ef}/f_0 \ll 1, \quad \omega_0 = 2\pi f_0, \tag{11.12}$$

where ω_0 is the average frequency of the signal spectrum. The inequality (11.12) for signals with a sinusoidal carrier means that a large number of oscillations of the high-frequency carrier occur on an interval of the order of the effective pulse duration, so that the condition of quasi-harmonicity may be written in an alternate form, as follows:

$$f_0 \vartheta_{ef} \gg 1. \tag{11.13}$$

From the inequality (11.13) we obtain the obvious integral estimate

$$\int_{-\infty}^{\infty} s_0^2(t)\, dt \gg \int_{-\infty}^{\infty} s_0^2(t) \cos \omega_0 t \, dt. \tag{11.14}$$

In studying the statistical characteristics of reverberation, the investigated signals turn out in almost every practical situation to be quasi-harmonic, and the inequalities cited above must hold true.

§ 12. Modulation Correlation Function

An important characteristic of the transmitted signals is their modulation correlation function, * which determines the statistical connections in the signal in terms of the frequency—time coordinates. By definition [3] it is equal to

$$C(\tau, \Omega) = (1/\delta_{ef}) \int_{-\infty}^{\infty} \overset{\ast}{s}_{01}(t)\, s_{01}(t+\tau) \exp(j\Omega t)\, dt, \tag{12.1}$$

where $s_{01}(t)$ is a function characterizing the laws of amplitude and frequency modulation of the signal and is defined by Eq.(10.4); $s_{01}^{\ast}(t)$ is the complex conjugate of the function $s_{01}(t)$; the effective pulse duration δ_{ef} enters into the expression (12.1) according to the conditions of normalization of the function $C(\tau, \Omega)$ at the point $\tau = 0$, $\Omega = 0$, so that

$$C(0, 0) = 1.$$

The modulation correlation function $C(\tau, \Omega)$ has the following basic properties:

The integral of $|C(\tau, \Omega)|^2$ over the variables τ and Ω is equal to 2π:

$$\int_{-\infty}^{\infty} \int_{-\infty}^{\infty} |C(\tau, \Omega)|^2 d\tau\, d\Omega = 2\pi, \tag{12.2}$$

which characterizes the invariance of the volume bounded by the surface $|C(\tau, \Omega)|^2$ and the time-frequency plane for any type of signal.

The function $C(\tau, \Omega)$ is expressed in terms of the spectral density

$$g_{01}(\omega) = \int_{-\infty}^{\infty} s_{01}(t) \exp(-j\omega t)\, dt \tag{12.3}$$

of the modulating function $s_{01}(t)$, as follows:

$$C(\tau, \Omega) = (1/2\pi\delta_{ef}) \int_{-\infty}^{\infty} g_{01}(\omega)\, g_{01}^{\ast}(\omega + \Omega) \exp(j\omega\tau)\, d\omega. \tag{12.4}$$

---

* This function is sometimes also called the signal autocorrelation function or the normalized ambiguity function.

The Fourier transform of $C(\tau, \Omega)$, i.e., the correlation modulation spectrum $G_C(\omega, \Omega)$, is defined by the relation

$$G_C(\omega, \Omega) = (1/\delta_{ef}) g_{01}(\omega) g_{01}^{*}(\omega + \Omega); \qquad (12.5)$$

this property follows directly from (12.4).

It is useful in some instances to apply such concepts as the time and frequency correlation intervals of the signal.

The effective time correlation interval $\tau_{ef}(\Omega)$, which is defined according to the expression

$$\tau_{ef}(\Omega) = \int\limits_{-\infty}^{\infty} |C(\tau, \Omega)|^2 d\tau, \qquad (12.6)$$

is equal to

$$\tau_{ef}(\Omega) = (1/2\pi) \int\limits_{-\infty}^{\infty} |G_C(\omega, \Omega)|^2 d\omega =$$
$$= (1/2\pi\delta_{ef}^{2}) \int\limits_{-\infty}^{\infty} |g_{01}(\omega)|^2 |g_{01}^{*}(\omega + \Omega)|^2 d\omega. \qquad (12.7)$$

For $\Omega = 0$ we have the relation

$$\tau_{ef}(0) = (1/\delta_{ef}^{2}) \int\limits_{-\infty}^{\infty} \left| \int\limits_{-\infty}^{\infty} s_{01}^{*}(t) s_{01}(t + \tau)\, dt \right|^2 d\tau. \qquad (12.8)$$

The effective frequency correlation interval $\Omega_{ef}(\tau)$, which is equal by definition to

$$\Omega_{ef}(\tau) = \int\limits_{-\infty}^{\infty} |C(\tau, \Omega)|^2 d\Omega, \qquad (12.9)$$

is expressed in terms of the spectrum $g_{01}(\omega)$ and the function $s_{01}(t)$ as follows:

$$\Omega_{ef}(\tau) = (1/4\pi^2\delta_{ef}^{2}) \int\limits_{-\infty}^{\infty} \left| \int\limits_{-\infty}^{\infty} g_{01}(\omega) g_{01}^{*}(\omega + \Omega) \exp(j\omega\tau)\, d\omega \right|^2 d\Omega =$$
$$= (2\pi/\delta_{ef}^{2}) \int\limits_{-\infty}^{\infty} |s_{01}^{*}(t) s_{01}(t + \tau)|^2 dt. \qquad (12.10)$$

That property of the function $C(\tau, \Omega)$ set forth in Eq. (12.2) reflects the so-called ambi-guity principle, whereby any deformations of the surface $C(\tau, \Omega)$ relative to the axes τ and Ω in connection with a variation in the form of modulation of the signal s(t) does not alter the volume bounded by the function $C(\tau, \Omega)$. This means, for example, that a contraction of $C(\tau, \Omega)$ in the direction of either the τ or the Ω axis will be accompanied by a corresponding elongation in the direction of the other coordinate.

Frequently, in order to simplify and render more transparent the graphical representation of $C(\tau, \Omega)$, so-called <u>ambiguity diagrams</u> are used. Ambiguity diagrams are usually understood to mean figures obtained in the coordinates τ, Ω by the intersection of the surface $|C(\tau, \Omega)|^2$ with the plane

$$|C(\tau, \Omega)|^2 = C_0^2, \tag{12.11}$$

where C_0 is a constant accounting for the reduction in correlation as a fraction of unity. In other words, for

$$|C(\tau, \Omega)|^2 < C_0^2$$

the signal correlation is negligible, while for

$$|C(\tau, \Omega)|^2 > C_0^2$$

it is arbitrarily agreed that the signal correlation is total.

The equation for the line of demarcation in this method of constructing the ambiguity diagrams is found from the condition (12.11).

In constructing the ambiguity diagrams, the following method, which does not involve arbitrariness in the choice of the constant C_0, is more valid in principle for the purpose of comparing the influence of the signal form and its modulation law on the form of the diagrams. The constant must be chosen so that the volume of a cylinder of unit height* having as its base the figure outlined in the plane (τ, Ω) by Eq.(12.11), will be equal to the ambiguity volume (12.2) for a signal of any form. This means that the level C_0 must be determined from the equation

$$\iint_{|C(\tau, \Omega)| = C_0} d\tau \, d\Omega = 2\pi. \tag{12.12}$$

We note that for certain types of signals the surface $|C(\tau, \Omega)|$ may have several "humps"; hence for each of the latter it acquires the interpretation of a local ambiguity diagram.

Consequently, the following procedure is used to find the ambiguity diagrams:

An arbitrary value of C_0 is specified somewhere on the interval $(0, 1)$.

The equation for the curve in the plane (τ, Ω) is determined according to the condition (12.11).

The area enclosed by the curve is found as a function of the parameter C_0.

The resultant area is equated to 2π in accordance with (12.12).

The constant C_0 is ascertained from the equation thus obtained.

The value of C_0 is substituted in Eq. (12.11), from which the ambiguity diagram is then constructed.

The indicated method of analyzing ambiguity diagrams is entirely general, because it reflects the condition of constant ambiguity volume for any signal form.

---

*In some cases the ambiguity function may be double-valued, in which case it is necessary to account for the total (summed) volume of both cylinders.

For some signals, however, the determination of the ambiguity diagrams by the procedure outlined above is a complicated computational task. It is necessary, therefore, to use the conventional method of constructing the diagrams on the basis of arbitrarily chosen values of the parameter C_0. It is sometimes the practice to construct a composite ambiguity diagram for several levels C_0, corresponding to regions of strong, medium, and weak correlation.

We conclude with some general expressions for the modulation correlation functions for signals with a sinusoidal carrier, frequency-modulated signals, and noisy signals.

For signals with a sinusoidal carrier $s_{01}(t) = s_0(t)$ and, consequently,

$$C(\tau, \Omega) = (1/\delta_{ef}) \int_{-\infty}^{\infty} s_0(t) s_0(t + \tau) \exp(j\Omega t) dt. \tag{12.13}$$

For frequency-modulated signals,

$$C(\tau, \Omega) = (1/\delta_{ef}) \int_{-\infty}^{\infty} s_0(t) s_0(t + \tau) \exp\{j[\Omega t + \Phi(t + \tau) - \Phi(t)]\} dt. \tag{12.14}$$

In the case of signals of the noise type (10.15) the modulating function $s_{01}(t)$ has the form

$$s_{01}(t) = s_0(t) E(t) \exp[j\psi(t)], \tag{12.15}$$

where $E(t)$ and $\psi(t)$ are stochastic processes. Consequently, in order to obtain the modulation correlation function, it is necessary to carry out statistical averaging of the product $s_{01}^*(t)$ · $s_{01}(t + \tau)$, which leads to the notion of the a v e r a g e   m o d u l a t i o n   c o r r e l a t i o n   f u n c - t i o n :

$$\langle C(\tau, \Omega) \rangle = (b_x(\tau)/\delta_{ef}) \int_{-\infty}^{\infty} s_0(t) s_0(t + \tau) \exp(j\Omega t) dt, \tag{12.16}$$

where

$$b_x(\tau) = \langle E(t) E(t + \tau) \exp\{j[\psi(t + \tau) - \psi(t)]\} \rangle. \tag{12.17}$$

If the process $x(t)$ generating the noise carrier of the signal has unit variance $\sigma_x^2 = 1$ and is invested with the property of quasi-harmonicity, Eq.(12.16) may be rewritten in the form

$$\langle C(\tau, \Omega) \rangle = (r_x(\tau)/\delta_{ef}) \int_{-\infty}^{\infty} s_0(t) s_0(t + \tau) \exp(j\Omega t) dt, \tag{12.18}$$

where $r_x(\tau)$ is the envelope of the correlation coefficient of the process and is found from the following representation of the correlation coefficient:

$$R_x(\tau) = r_x(\tau) \cos \omega_0 \tau. \tag{12.19}$$

CHAPTER III

PROBABILITY DISTRIBUTIONS OF
REVERBERATION SIGNALS

§ 13. Limiting Distributions of the Instantaneous Values

The reverberation signals described by some stochastic time function F(t), according to the statistical model of sound scattering adopted in the first chapter, represent a summation of a random number of elementary scattered signals arriving simultaneously at the point of reception and having random amplitudes and onset times.

We first investigate the characteristics of the stationary part V(t) of reverberation. The stochastic function V(t) is a part of F(t) and is found from the representation of F(t) in the form (9.3).

If the number of scattered signals $\langle n \rangle \delta_{ef}$ arriving simultaneously at the reception point at given instants of time is large, i.e., if the following inequality holds:

$$\langle n_1 \rangle \delta_{ef} \gg 1, \tag{13.1}$$

where δ_{ef} is the effective duration of the transmitted pulses, then the one-dimensional distribution W(V) will be described by a normal law, in view of the central limit theorem, provided sufficiently general conditions are satisfied with a definite degree of accuracy.

$$W(V) = \left(1/\sqrt{2\pi}\sigma_V\right) \exp\left(-V^2/2\sigma_V^2\right), \tag{13.2}$$

where σ_V^2 is the reverberation variance. The mean value of the reverberation is zero, since the transmitted signals are quasi-harmonic.

We now state the conditions under which the process V(t) will be exactly described by a normal law. A normal distribution is valid in the limit as $\langle n_1 \rangle \delta_{ef} \to \infty$ for any transmitted pulse form and any type of elementary amplitude distribution of the elementary scattered signals, provided none of these signals is predominant over the rest and hence does not yield a significant contribution to the total process.

In this sense the inequality (13.1) does not in general lead to the limiting relation (13.2), and in the ensuing sections we will examine the problems of the convergence of the reverberation distribution to a normal law.

It is also admissible for the stationary process V(t) to write a multidimensional normal distribution $W(V_1, t_1; V_2, t_2; \ldots; V_n, t_n) = W(V_1, V_2, \ldots, V_n)$ of a set of n values $V(t_1), V(t_2), \ldots, V(t_n)$.

This distribution has the following form (see, e.g., [10]):

$$W\left(V_1, V_2, \ldots, V_n\right) = \frac{1}{(2\pi)^{n/2}\sigma_V^n |K|^{1/2}} \exp\left[\left(-\tfrac{1}{2}|K|\sigma_V^2\right)\sum_{i=1}^{n}\sum_{m=1}^{n}|K_{im}|V_iV_m\right], \tag{13.3}$$

where K is the determinant of the correlation matrix:

$$K = \begin{Vmatrix} R_{11} & R_{12} & \ldots & R_{1n} \\ R_{21} & R_{22} & \ldots & R_{2n} \\ \cdot & \cdot & \cdot & \cdot \\ R_{n1} & R_{n2} & \ldots & R_{nn} \end{Vmatrix}, \tag{13.4}$$

$|K_{im}|$ is the signed minor of the element R_{im} in the determinant of the matrix (13.4). The elements R_{im} of the matrix are the correlation coefficients and are equal to

$$R_{im} = \langle V(t_i)V(t_m)\rangle/\sigma_V^2; \quad R_{ii} = 1; \quad R_{im} = R_{mi}; \tag{13.5}$$

taking into account the stationarity of the process V(t), we write for R_{im}

$$R_{im} = R(t_m - t_i) = \langle V(t_i)V(t_m)\rangle/\sigma_V^2. \tag{13.6}$$

We note that the stationarity of reverberation signals is observed despite the variation of the parameter $<n_1> \delta_{\text{ef}}$ over wide limits. But if the condition (13.1) is not fulfilled or is only scarcely fulfilled, the process will be, first of all, nonstationary and irreducible to stationary and, secondly, the set of values $V(t_1), V(t_2), \ldots, V(t_n)$ will not fit the n-dimensional normal distribution (3.13).

If the variances $\sigma_{V_1}, \sigma_{V_2}, \ldots, \sigma_{V_n}$ corresponding to the different times $t_1, t_2, \ldots, t_n$ are not equal, but the process V(t) is reducible to a stationary normal process, then

$$W\left(V_1, V_2, \ldots, V_n\right) = \frac{1}{(2\pi)^{n/2}|K|^{1/2}\prod\limits_{i=1}^{n}\sigma_{V_i}} \exp\left[\left(-\tfrac{1}{2}|K|\right)\sum_{i=1}^{n}\sum_{m=1}^{n}\frac{|K_{im}|V_iV_m}{\sigma_{V_i}\sigma_{V_m}}\right], \tag{13.7}$$

where

$$\sigma_{V_1} = \langle V^2(t_1)\rangle, \; \sigma_{V_2} = \langle V^2(t_2)\rangle, \ldots, \sigma_{V_n} = \langle V^2(t_n)\rangle. \tag{13.8}$$

The correlation coefficients R_{im} in this case are defined by the relations

$$R_{im} = R(t_m - t_i) = \langle V(t_i)V(t_m)\rangle/\sigma_i\sigma_m, \quad i, m = 1, 2, \ldots, n. \tag{13.9}$$

The averaging process by which the statistical characteristics of reverberation, including the distribution (13.7), can be found must be performed in this case over the set of states.

It is interesting to note that for $i \neq m$ also the correlation coefficient determined according to (13.9) will not depend on the time. This condition is then a criterion of whether the reverberation is attributable to processes that are reducible to stationary (see Sec. 9).

It was postulated above that the amplitudes of the elementary scattered signals obey some rather arbitrary distribution law W(a).

However, in the observation of reverberation under real conditions, one or more scattered signals may arrive at the reception point at some instant in time with amplitudes comparable with the mean-square value of the reverberation or even exceeding it. The influence of such signals is best treated separately, assuming that the amplitudes of the remaining elementary scattered signals are described by the distribution law W(a).

Below we consider the one-dimensional distributions of reverberation for such a case.

Let the transmitted pulse represent a segment of a harmonic vibration of frequency ω_0 with a rectangular envelope, so that for instants when a fairly intense scattered signal with constant amplitude b_0 is present the reverberation $V_1(t)$ may be represented in the form

$$V_1(t) = V(t) + b(t) = V(t) + b_0 \cos(\omega_0 t + \varphi), \tag{13.10}$$

where V(t) is the normally distributed component of the reverberation, φ is the random initial phase for which, in view of the fluctuations in the conditions of sound propagation, motion of the scatterers, and other factors, a uniform distribution law is assumed:

$$W(\varphi) = 1/2\pi, \quad |\varphi| \leqslant \pi. \tag{13.11}$$

The probability density W(V_1) may be found, assuming statistical independence of the processes V(t) and b(t), either by the method of characteristic functions or by the convolution of the functions W(V) and W(b). Consequently, this problem reduces to a determination of the probability density of the sum of a normal process and a harmonic signal with random initial phase [1, 12, 51].

The probability density W(b) of the process b(t) is defined by the expression

$$W(b) = \frac{1}{\pi b_0 \sqrt{1 - (b/b_0)^2}}, \quad |b/b_0| \leqslant 1. \tag{13.12}$$

The convolution of W(V) and W(b) with application of (13.2) and (13.12) leads to the following relation for W(V_1):

$$W(V_1) = \frac{1}{\sqrt{2\pi^3 b_0 \sigma_V}} \int_{-b_0}^{b_0} \exp\left[-(V_1 - x)^2/2\sigma_V^2\right] \frac{dx}{\sqrt{1 - (x/b_0)^2}}$$

or, after suitable replacement of the variable of integration, to the equivalent relation

$$W(V_1) = \frac{1}{\sqrt{2\pi^3 b_0 \sigma_V}} \int_{-\pi}^{\pi} \exp\left[-\frac{(V_1 - b_0 \cos\theta)^2}{2\sigma_V^2}\right] d\theta. \tag{13.13}$$

For calculations of the expressions for the distribution (13.13) the expansion of W(V_1) in series of Bessel functions $I_k(X)$ of an imaginary argument, in confluent hypergeometric functions $_1F_1(X, Y, Z)$, or in the derivatives $\varphi^{(k)}(X)$ of a square exponential, proves more suitable.

These expressions have the following form (see, e.g., [12]):

$$W(V_1) = (1/\sqrt{2\pi}\sigma_V)\exp\left[-(2V_1^2 + b_0^2)/4\sigma_V^2\right]\left\{I_0(V_1 b_0/\sigma_V^2) I_0[b_0^2/4\sigma_V^2] + \right.$$
$$\left. + 2\sum_{k=1}^{\infty} (-1)^k I_k[b_0^2/4\sigma_V^2] I_{2k}(V_1 b_0/\sigma_V^2)\right\}, \tag{13.14}$$

$$W(V_1) = (1/\sqrt{2\pi}\sigma_V)\sum_{k=0}^{\infty}(-V_1/2\sigma_V^2)^k (1/k!)_1F_1[(k + 1/2), 1, -b_0^2/2\sigma_V^2], \tag{13.15}$$

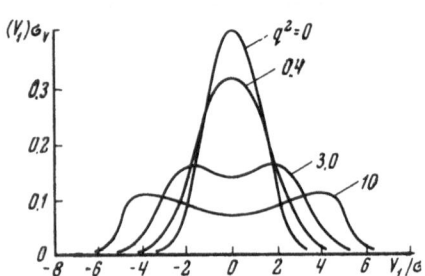

Fig. 14. Distributions of the instantaneous values of the reverberation in the presence of a determinant scattered signal of constant amplitude.

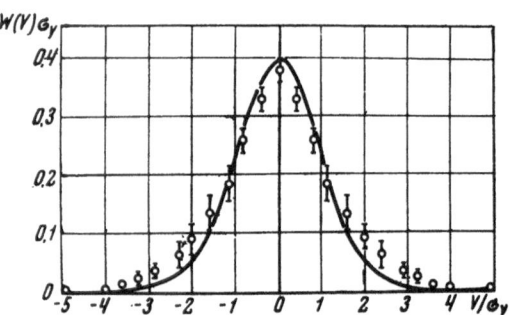

Fig. 15. Comparison of a normal distribution with the experimental data (dots) on the distribution of the instantaneous reverberation values.

$$W(V_1) = (1/\sigma_V) \sum_{k=0}^{\infty} (1/2^k) [1/(k!)^2] (b_0/\sigma_V)^{2k} \varphi^{(2k)} (V_1/\sigma_V), \qquad (13.16)$$

$$\varphi^{(k)}(x) = \frac{d^k}{dx^k} [(1/\sqrt{2\pi}) \exp(-x^2/2)].$$

Graphs of the normalized probability density $W(V_1)\sigma_V$ are shown in Fig. 14 for various values of the parameter,

$$q^2 = \frac{b_0^2}{2\sigma_V^2}, \qquad (13.17)$$

which characterizes the ratio of the intensity of the individual scattered signal to the variance of the reverberation process $V(t)$.

As apparent from the graph, the probability density for any values of the parameter q is symmetrical relative to $V_1/\sigma_V = 0$ and degenerates into a double-humped curve for large values of that parameter.

Appropriate experimental investigations of the one-dimensional distributions have been carried out for the purpose of confirming the validity of the above relations in application to the analysis of reverberation signals.

Data from measurements of the distribution of instantaneous values of the combined surface and bottom reverberation are shown in Fig. 15 by way of example. The reverberation was processed by means of a type AI-100 instrument for the analysis of more than 100 independent readings of the reverberation process in each series of measurements. The averaged data (dots) represent the result of processing five series of measurements. Figure 15 clearly indicates the presence of a regular component in the reverberation process due to scattering by inhomogeneities of the ocean bottom, which also resulted in a small negative excess. The observed excess corresponds to a value of q ≈ 0.3.

We have thus investigated the influence of an intense scattered signal on the form of the reverberation distribution for the transmission of pulses with a sinusoidal carrier and rectangular envelope.

It is interesting to note that the distribution (13.13) must also be valid in the presence of coherent scattering.

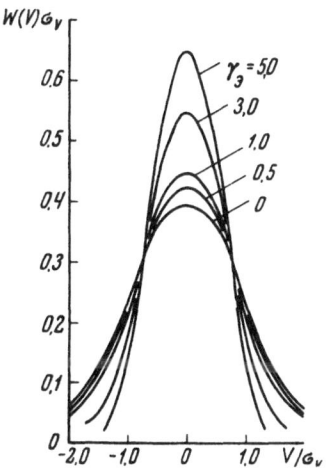

Fig. 16. Distribution of instantaneous reverberation values for various values of the coefficient of excess.

The coefficient q, defined according to (13.17), has the sense here of the ratio of the coherent, i.e., the determinate, component to the purely stochastic component, for which a normal distribution law is assumed valid. If q depends on the time, i.e., if in the reverberation process the ratio of the coherent and stochastic component varies, the function $V_1(t)$ is highly nonstationary and it is no longer possible in this case to speak of reverberation as a process reducible to stationary even in the broad sense.

We also indicate one feature of the reverberation distribution for the transmission of noisy pulses. The reverberation $V_1(t)$ at the time of arrival of an intense scattered signal represents the sum

$$V_1(t) = V(t) + b(t). \tag{13.18}$$

If b(t) is a normal stochastic process with variance σ_b^2, then, for the reverberation distribution, a normal law with the following variance is also valid:

$$\sigma_{V_1}^2 = \sigma_V^2 + \sigma_b^2. \tag{13.19}$$

Here the probability density $W(V_1)$, taking (13.19) into account, is determined by Eq. (13.2) on the interval $(t - \delta_{ef}/2, t + \delta_{ef}/2)$ for which $b(t) \neq 0$. The multidimensional distribution obtained in the indicated time interval by averaging over the ensemble of states will be described by a relation of the form (13.3), while for an arbitrary set of times the more general equation (13.7) is applicable for the n-dimensional distribution.

§ 14. Distributions for a Finite Number of Elementary Scattered Signals

In the preceding section we investigated the limiting distributions of the instantaneous reverberation values with an increase in the average number of elementary scattered signals arriving simultaneously at the point of reception.

We now assess the influence of the parameter $<n_1>\delta_{ef}$ on the one-dimensional distribution function W(V) if the value of this parameter is small.

Such cases may be encountered, for example, in observing bottom reverberation at small distances, volume reverberation due to scattering by a small number of fairly localized inhomogeneities, etc.

In fact, at relatively short distances, particularly with the use of acoustic arrays having narrow directivity patterns and in the transmission of pulses of short duration, the scattering volume may prove to be quite small. If, accordingly, the mean spatial density of the scatterers in the ocean is relatively small, the average number of elementary scattered signals generating the reverberation process will also be small.

We ascertain the distributions of the instantaneous reverberation values by an approximate method using a series representation of the probability density. Inasmuch as the limiting distribution as $<n_1>\delta_{ef} \to \infty$ is normal in this case, a convenient representation of W(V) is the expansion of the probability density, say, in the series (8.11) or the approximate series (8.13).

If the transmitted signals are quasi-harmonic, all of the odd-order semi-invariants will be equal to zero, in particular $<V> = 0$, $\gamma_a = 0$.

In this case, Eq. (8.13) takes the form

$$W(V) \approx \left(1/\sqrt{2\pi}\sigma_V\right)\exp\left(-V^2/2\sigma_V^2\right)\ [1 + (\gamma_e/4!)\,H_4(V)], \tag{14.1}$$

where γ_e is the coefficient of excess defined in (6.30).

Making use of (8.5) and (14.1), we obtain for the probability density W(V)

$$W(V) \approx \left(1/\sqrt{2\pi}\sigma_V\right)\exp\left(-V^2/2\sigma_V^2\right)\ \{1 + (\gamma_e/4!)\,[(V/\sigma_V)^4 - 6\,(V/\sigma_V)^2 + 3]\}. \tag{14.2}$$

It follows from this relation that as $\gamma_e \to 0$ the distribution W(V) converges to a normal distribution.

For $\gamma_e > 0$, the reverberation distribution departs from a normal law and has a positive excess.

Normalized probability densities $W(V)\sigma_V$ computed according to Eq. (14.2) are shown in Fig. 16. It is apparent from the graph that the distribution of instantaneous reverberation values practically coincide with a normal distribution for $\gamma_e < 0.5$ and for values of V/σ_V in the interval $(-2, 2)$ do not depart from it by more than 5-7% on the average.

It is important to realize that for large values of γ_e, above 7-10, the accuracy ensured by Eq. (14.2) is not adequate, and it is necessary for the purpose of the calculations to choose a larger number of terms in the original series (8.11).

The value of the coefficient of excess depends in general on the form of the transmitted pulses, the average number of scattered signals arriving simultaneously at the reception point, and their amplitude distribution. We discuss these dependences in the next section.

§ 15. Coefficient of Excess

We now investigate the influence of the form of the transmitted pulses, the amplitude distribution of the elementary scattered signals, and their average number on the coefficient of excess for the distribution of the instantaneous reverberation values. It is apparent from Eq. (14.2) that the coefficient of excess γ_e determines the form of the probability density curve for the reverberation and characterizes its degree of proximity to a normal distribution.

Inasmuch as γ_e is expressed in terms of the second- and fourth-order semi-invariants, which may be found by means of the one-dimensional theorem of superposition of stochastic perturbations, we obtain, on the basis of (5.4) and (6.30),

$$\gamma_e = \frac{\langle a^4 \rangle \int\limits_{-\infty}^{\infty} s^4(t)\,dt}{\langle n_1 \rangle \langle a^2 \rangle^2 \left[\int\limits_{-\infty}^{\infty} s^2(t)\,dt\right]^2}. \tag{15.1}$$

We next introduce certain parameters to characterize the effect of the scattered signal amplitudes, their average number, and the form of the transmitted pulses on the coefficient of excess. Such parameters include, for example,

$$A = \langle a^4 \rangle/\langle a^2 \rangle^2, \tag{15.2}$$

Table 15.1.   Typical Distributions of the Amplitudes of Elementary
Scattered Signals and Their Parameters

No.	Type of distribution W_a	$\langle a^2 \rangle$	$\langle a^4 \rangle$	A
1	Exponential $(1/a_0) \exp(-a/a_0), \quad a \geqslant 0$	$2a_0^2$	$24a_0^4$	6
2	Rayleigh $(a/a_0^2) \exp(-a^2/2a_0^2), \quad a \geqslant 0$	$2a_0^2$	$8a_0^4$	2
3	Uniform $1/a_0, \quad 0 \leqslant a < a_0$	$a_0^2/3$	$a_0^4/5$	1.8
4	Delta function (case of constant amplitudes) $\delta(a - a_0), \quad a > 0$	a_0^2	a_0^4	1

$$\langle n \rangle = \langle n_1 \rangle \delta_{\text{ef}}, \tag{15.3}$$

and

$$S = \frac{\delta_{\text{ef}} \int\limits_{-\infty}^{\infty} s^4(t)\, dt}{\left[\int\limits_{-\infty}^{\infty} s^2(t)\, dt \right]^2}. \tag{15.4}$$

Bearing in mind the definitions (15.2)-(15.4), we go from the relation (15.1) to the following:

$$\gamma_e = AS/\langle n \rangle. \tag{15.5}$$

Hence it is apparent that with an increase in $<n>$ and a reduction in the parameters A and S, the value of γ_e decreases.   For $<n> \to \infty$ in particular, the coefficient of excess tends to zero, i.e., we arrive at a normal distribution for W(V), which follows at once from (14.2).

Below we examine certain types of distributions W(a) of the elementary scattered signal amplitudes, taking in the main characteristic conditions of sound scattering by various inhomogeneities of the ocean medium.   These distributions are summarized in Table 15.1, which also gives the values of the moments $<a^4>$ and $<a^2>$ and the parameter A.   In choosing the distributions, due consideration was given the fact that scattered signals with excessively small amplitudes could not significantly affect the value of the parameter A, while intense signals must be accounted for separately.

We point out that the value of the parameter a_0 occurring in the expressions for the distributions shown in the table do not influence the coefficient A, since the latter is a normalized characteristic and depends only on the type of probability density W(a).

It is evident from the data of Table 15.1 that the parameter A has its minimum value in the case of the distribution described by a delta function, for which the amplitudes of all the scattered signals are equal.   An exponential distribution yields the largest value of A of all the cases considered; this is entirely reasonable, considering that equal amplitudes prove extremely

Table 15.2. Values of the Parameter S for Pulses with Various Envelopes

No.	Type of pulse envelope $S_0(t)$	S
1	Rectangular $1,\ |t| \leqslant \delta/2$	1.50
2	Bell-shaped $\exp[-(t/t_0)^2]$	1.06
3	Type $(\sin x/x)$ $\dfrac{\sin(t/t_0)}{t/t_0}$	1.00
4	Exponential $\exp(-t/t_0),\ t \geqslant 0$	0.75

nonequivalent for an exponential distribution, a fact which does not promote normalization of the distribution of reverberation signals.

We next consider the influence of the form of transmitted signals having a sinusoidal carrier and envelopes of various shapes on the coefficient of excess. In the general case such pulses are written in the form (10.6). Under the condition of quasi-harmonicity of the function S(t), the parameter S is determined as follows, taking (11.1) and (15.4) into account:

$$S = (1.5/\delta_{\text{ef}}) \int_{-\infty}^{\infty} s_0^4(t)\, dt. \qquad (15.6)$$

The results of calculating this parameter are shown in Table 15.2 for the four types of pulses most pertinent to the investigation of the statistical characteristics of reverberation.

From an analysis of the parameters A, S, and $<n>$ and their influence on the coefficients of excess, we can draw some inference as to the speed of convergence of the reverberation distribution to a normal distribution as a function of the form of the elementary scattered signals and their amplitude distributions.

It is significant that the most rapid convergence to a normal law (for $<n>$ = const) is observed for equal scattered-signal amplitudes (see row 4 of Table 15.1) and an exponential transmitted-signal envelope (see row 4 of Table 15.2). Conversely, the poorest reverberation distribution as far as convergence to a normal law occurs for an exponential amplitude distribution (see row 1 of Table 15.1) and a rectangular envelope (see row 1 of Table 15.2).

We now consider the case of amplitude-modulated transmitted signals. Consistent with (10.7), (10.10), and (15.6) we have for the parameter S

$$S = \frac{1.5\,(1+m)^2\,(1+3m^2+3m^4/8)}{1+m^2/2}. \qquad (15.7)$$

It follows from Eq. (15.7) that as the modulation index is increased the parameter S, and hence the coefficient of excess γ_e, increase; thus, in going from m = 0 (no modulation) to m = 1 (total modulation), the value of γ_3 increases approximately twelve-fold.

We conclude by pointing out one feature of the convergence of the reverberation distribution to a normal law in the transmission of noisy signals. When the carrier is noise having a normal distribution, the reverberation distribution will also be normal for any values of $<n>$ and A. This is true by virtue of the fact that the sum of any number of normally distributed components obeys a normal distribution law.

It is possible, proceeding in analogous fashion, to calculate the coefficients of excess and to evaluate the degree of proximity of the reverberation distribution to a normal law for other forms of transmitted signals.

§16. Quadrature Components of Reverberation and the Hilbert Transform

The reverberation process V(t), like any quasi-harmonic stochastic function, may be written in the form

$$V(t) = E(t) \cos [\omega_0 t + \psi(t)], \tag{16.1}$$

where E(t) is the envelope, $\psi(t)$ is the phase of the reverberation, ω_0 is the central frequency of the spectrum. It is known from the theory of quasi-harmonic processes (see [1, 12, 19]) that E(t) and $\psi(t)$ are rather slowly varying functions of the time in comparison with $\cos \omega_0 t$.

It is advantageous for the purpose of later analysis to rewrite Eq. (16.1) in the form

$$V(t) = V_c(t) \cos \omega_0 t - V_s(t) \sin \omega_0 t, \tag{16.2}$$

where the stochastic processes $V_C(t)$ and $V_S(t)$ are defined by

$$\left.\begin{array}{l} V_c(t) = E(t) \cos \psi(t), \\ V_s(t) = E(t) \sin \psi(t) \end{array}\right\} \tag{16.3}$$

and represent the quadrature components of reverberation.

It is apparent that for the envelope and phase of the reverberation we may write, on the basis of (16.3),

$$\left.\begin{array}{l} E(t) = [V_c^2(t) + V_s^2(t)]^{1/2} \\ \psi(t) = \text{arc tg } [V_s(t)/V_c(t)] \end{array}\right\}. \tag{16.4}$$

The quadrature components of reverberation $V_C(t)$ and $V_S(t)$ as stochastic processes have the following fundamental properties when certain general conditions are met:

For reverberation with a normal distribution, these processes also have normal distributions with zero mean values and variances $\sigma_{V_c}^2$ and $\sigma_{V_s}^2$ equal to the variance σ_V^2 of the reverberation, i.e.,

$$\left.\begin{array}{l} W(V_c) = \left(1/\sqrt{2\pi}\sigma_{V_c}\right) \exp\left(-V_c^2/2\sigma_{V_c}^2\right), \\ W(V_s) = \left(1/\sqrt{2\pi}\sigma_{V_s}\right) \exp\left(-V_s^2/2\sigma_{V_s}^2\right), \end{array}\right\} \tag{16.5}$$

$$\sigma_{V_c}^2 = \sigma_{V_s}^2 = \sigma_V^2; \tag{16.6}$$

The quadrature components are uncorrelated and, considering that the multidimensional distribution of reverberation is normal (see Sec. 13), statistically independent; consequently,

$$\left.\begin{array}{l} \langle V_c(t) V_s(t) \rangle = 0, \\ W(V_c, V_s) = W(V_c) W(V_s); \end{array}\right\} \tag{16.7}$$

The quadrature components $V_C(t)$ and $V_S(t)$ are interrelated by the Hilbert transform (see, e.g., [3, 12, 19]):

$$V_s(t) = -(1/\pi) \lim_{T \to \infty} \int_{-T}^{T} \frac{V_c(t') dt'}{t - t'}, \tag{16.8}$$

$$V_c(t) = (1/\pi) \lim_{T \to \infty} \int_{-T}^{T} \frac{V_s(t')\,dt'}{t - t'}$$

and are sometimes called c o n j u g a t e processes.

We note that in general the Hilbert transform

$$y(t) = H[x(t)] = -(1/\pi) P \int_{-\infty}^{\infty} \frac{x(t')\,dt'}{t - t'}$$

is understood to mean the principle value of the integral, as indicated by the index P.

The Hilbert transform therefore represents a linear integral operation with kernel

$$h(t-t') = 1/\pi(t-t').$$ (16.9)

It has been shown (see [19]) that the Hilbert transform, which essentially rotates the phases of all the components of the frequency spectrum for the investigated stochastic process through an angle $\pi/2$, cannot be exactly realized by means of physically tenable linear systems.

Hence, it is possible to find a frequency characteristic K(ω) corresponding to the kernel (16.9). Since

$$K(\omega) = \int_{-\infty}^{\infty} h(x) \exp(-j\omega x)\,dx,$$

we have, with allowance for (16.9),

$$K(\omega) = (1/\pi) \int_{-\infty}^{\infty} \frac{\exp[-j\omega(t-t')]}{t-t'}dt' = (1/\pi) \int_{-\infty}^{\infty} \frac{\exp(-j\omega x)}{x}\,dx =$$

$$= (1/\pi) \int_{0}^{\infty} \frac{\exp(-j\omega x) - \exp(j\omega x)}{x}\,dx.$$

For real values of the frequency ω, making use of the relation

$$(2/\pi) \int_{0}^{\infty} \frac{\sin(\omega x)\,dx}{x} = \begin{cases} 1, & \omega > 0, \\ -1, & \omega < 0, \end{cases}$$

we obtain for the frequency characteristic K(ω)

$$K(\omega) = \begin{cases} -j, & \omega > 0, \\ j, & \omega < 0. \end{cases}$$ (16.10)

This equation shows that a linear system by means of which it is possible to realize the Hilbert transform of the initial function does not affect the amplitude of that function; it only changes the phase of each of its frequency components by an amount $\pi/2$.

In the case of sufficiently narrow-band processes or, more precisely, processes with a decaying spectrum, this transform can be approximately realized by means of analog systems. Moreover, once the two quadrature components have been obtained, it is possible by application, for example, of the techniques of computer technology, to find the reverberation envelope and phase according to the algorithm (16.4).

This is specifically the method used in [31] to investigate certain properties of the reverberation envelope and phase; this paper presents the results of an investigation of the quadrature components, as the Hilbert transform, on a digital computer. The Hilbert transforms were obtained by means of a special analog system, a wide-band phase shifter; the reverberation signals were fed to the input of the device, which was then used in a fairly broad frequency band (sonic range) to accomplish a frequency-independent phase rotation of the frequency components through $\pi/2$. Then the algorithm (16.4) used to discriminate the reverberation envelope E(t) and phase ψ(t) (interpreted in the strict sense) was realized on the digital computer.

It is important to note that for stochastic processes representable by a relation of the type (16.1), all the information concerning them is contained in the processes E(t) and ψ(t); hence, the stochastic process under investigation is completely defined by specification of those two processes. Here, generally speaking, the initial process V(t) is not necessarily quasi-harmonic, because the representation

$$V(t) = E(t) \cos \Phi(t)$$

has an extremely general interpretation and can, in general, be associated with any process; but the case when

$$\Phi(t) = \omega_0 t + \psi(t)$$

leads to the representation (16.1), which corresponds to the quasi-harmonic process V(t).

§17. Instantaneous Phase Distribution of Reverberation

We now examine certain statistical characteristics of the instantaneous phase ψ(t) determined according to the relation (16.4) in terms of the quadrature components of reverberation V_c(t) and V_s(t).

If the number of elementary scattered signals arriving simultaneously at the reception point and generating the reverberation process is very large, the multidimensional distribution (13.3) holds for the stationary component of reverberation. The stochastic process V(t) in this case turns out to be ergodic. The following assertion applies to such processes:

If V(t) is a stationary process, the one-dimensional probability density W(ψ) of the phase has a uniform distribution in the interval $(0, 2\pi)$,* i.e.,

$$W(\psi) = 1/2\pi, \qquad 0 \leqslant \psi < 2\pi. \tag{17.1}$$

The above statement is consistent with the physical notions of sound scattering by inhomogeneities of the ocean medium and the model of reverberation as a stochastic process. In general, it is not assumed here that V(t) is a normal process, and the formulated theorem is valid for a broader class of processes than just normal processes.

In fact, at any instant of time a set of elementary scattered signals arrives at the reception point with random onset times, i.e., random initial phases. The absence of an ordered spatial distribution of scatterers in the medium leads to the logical conclusion, which ensues, for example from the vectorial representation of reverberation, that its phase must obey a dis-

---

*Apropos of this problem, see, for example, the following article by F.V. Bunkin and L.I. Gudzenko: "The one-dimensional amplitude and phase distributions of a stationary process," Radiotekhnika i elektronika, 3(7):968-969 (1957). In this paper, time independence of the one-dimensional probability density W(E) of the envelope and conditional probability density W(ψ/E) of the phase is indicated as a sufficient condition for the validity of Eq.(17.1).

tribution of the type (17.1). A uniform distribution of the instantaneous reverberation phase may also be deduced from the Poisson law (4.2) for the number of elementary scattered signals. In particular, it turns out here that a uniform distribution is valid for the arrival times of these signals at the reception point, whence a uniform distribution of the instantaneous reverberation phase is immediately inferred.

Violation of the law (17.1) could be attributable either to the presence of intense scattered signals at certain instants of time or to the onset of coherent scattering.

Let us examine further some of the statistical characteristics of the instantaneous phase of reverberation, when regular signals of the following type are present at the reception point:

$$b\,(t) = b_0\,\cos\,(\omega_0 t + \varphi_0), \tag{17.2}$$

where b_0 is the constant amplitude, φ_0 is the initial phase, which we will set equal to zero for simplicity. In this case, the total reverberation process $V_1(t)$ is written in the form

$$V_1\,(t) = V\,(t) + b\,(t).$$

Considering that $V(t)$ is defined by Eq. (16.2), we have

$$V_1(t) = [V_c\,(t) + b_0]\cos\omega_0 t - V_s\,(t)\sin\omega_0 t = V_{c1}\,(t)\cos\omega_0 t - V_{s1}\,(t)\sin\omega_0 t. \tag{17.3}$$

Clearly, for $V_1(t)$, as for any quasi-harmonic process, a representation similar to (16.1) is valid, i.e.,

$$V_1(t) = E_1(t)\,\cos\,[\omega_0 t + \psi_1(t)], \tag{17.4}$$

where $E_1(t)$ and $\psi_1(t)$ are the envelope and phase of the total process $V_1(t)$.

The problem of analyzing the distribution $W(\psi_1)$ of the phase $\psi_1(t)$ therefore reduces to a calculation of the phase probability density for the combined normal stochastic process $V(t)$ and a sinusoidal signal of constant amplitude (see [1, 12]).

Let us analyze some of the statistical characteristics of the phase $\psi_1(t)$.

Proceeding from (17.3) and (17.4), we write the following relations for $E_1(t)$ and $\psi_1(t)$:

$$\left.\begin{array}{l} E_1\,(t) = \{[V_c\,(t) + b_0]^2 + V_s\,(t)\}^{1/2}, \\ \psi_1\,(t) = \tan^{-1}\,[V_s\,(t)/V_c\,(t) + b_0]. \end{array}\right\} \tag{17.5}$$

The probability density $W(\psi_1)$ is found on the basis of the general expression

$$W\,(\psi_1) = \int_0^\infty E_1 W_V\,(E_1\cos\psi_1,\ E_1\sin\varphi_1)dE_1,\ 0\leqslant\psi_1 < 2\pi, \tag{17.6}$$

where $W_V(V_{c1}, V_{s1})$ is the two-dimensional probability distribution of the processes $V_{c1}(t)$ and $V_{s1}(t)$ determined according to (17.3).

We have for $W_V(V_{c1}, V_{s1})$, by virtue of Eqs. (16.5)-(16.7),

$$W_V\,(V_{c1},\ V_{s1}) = (1/2\pi\sigma_V^2)\exp\,\{-[(V_{c1} - b_0)^2 + V_{s1}^2]/2\sigma_V^2\}$$

and, hence,

$$W_V\,(E_1\cos\psi_1,\ E_1\sin\psi_1) = (1/2\pi\sigma_V^2)\exp\,\{-[(E_1\cos\psi_1 - b_0)^2 + E_1^2\sin^2\psi_1]/2\sigma_V^2\}. \tag{17.7}$$

Moreover, substituting (17.7) into (17.6) and making a few elementary transformations, we obtain

$$W(\psi_1) = (1/2\pi\sigma_V^2) \exp(-b_0^2/2\sigma_V^2) \int_0^\infty E_1 \exp[-(E_1^2 - 2E_1 b_0 \cos\psi_1)/2\sigma_V^2] dE_1.$$

The integration yields the following relation for the unknown probability density of the reverberation phase:

$$W(\psi_1) = (1/2\pi) \exp(-q^2) + (q/2\sqrt{\pi}) \cos\psi_1 \exp(-q^2 \sin^2\psi_1)[1 + \Phi(q\cos\psi_1)], \qquad (17.8)$$

where, as in Sec.13, the parameter $q = b_0/\sqrt{2\sigma_V}$ characterizes the ratio of the effective amplitude of the regular signal to the mean-square value of the stochastic component of the reverberation, $\Phi(X)$ is the probability integral defined by Eq.(11.9).

We now indicate certain features of the resultant reverberation phase distribution for various values of the parameter q. For $q \ll 1$, i.e., for small relative values of b_0, we have a nearly uniform distribution.

It is interesting to note that if we let q = 0, the uniform distribution (17.1) follows for the instantaneous phase directly from Eq.(17.8). Consequently, two postulates are sufficient in order for (17.1) to be valid, to wit:

The pertinent values of the phase are determined in the interval $(0, 2\pi)$.

The quadrature components of the reverberation are statistically independent and obey a normal distribution law.

For $q \gg 1$, i.e., when the amplitude of the regular signal b_0 is large in comparison with σ_V, a normal distribution is approximately valid for $W(\psi_1)$, namely:

$$W(\psi_1) \approx (1/\sqrt{2\pi}\sigma_{\psi_1}) \exp(-\psi_1^2/2\sigma_{\psi_1}^2), \qquad (17.9)$$

where

$$\sigma_{\psi_1}^2 = 1/2q^2 = \sigma_V^2/b_0^2 \qquad (17.10)$$

is the variance of the reverberation phase fluctuations. We note that in this case the reverberation phase fluctuations are generally insignificant, and

$$\sigma_{\psi_1} \ll 2\pi.$$

The above results are applicable when the transmitted signals have a rectangular envelope, i.e., when $b_0 = $ const. Then, on the interval $(t - \delta/2, t + \delta/2)$ of existence of a signal to which these results are applicable, the process $V_1(t)$ may be assumed stationary. If, however, the transmitted signals have a more complicated envelope or a particular form of frequency modulation, the process $V_1(t)$ turns out to be nonstationary, and it is only admissible to speak of a probability density $W(\psi_1, t)$ depending on the running time t. The averaging under such conditions must be carried out over the ensemble of states of the investigated process.

§18. Envelope Distributions

The reverberation process was represented above in the form (16.1); such a representation permits the reverberation envelope distributions to be found when certain initial assumptions are given.

We first investigate the case when the initial process V(t) is described by a normal distribution and the values of the reverberation phase are considered on the interval $(0, 2\pi)$.

The following general relation holds for the one-dimensional probability distribution W(E) of the reverberation envelope:

$$W(E) = E \int_0^{2\pi} W_V(E \cos \psi, \ E \sin \psi) \, d\psi, \ E \geqslant 0, \tag{18.1}$$

where, as in Sec. 17, $W_V(V_c, V_s)$ is the two-dimensional probability density for the quadrature components of the reverberation $V_c(t)$ and $V_s(t)$. Taking Eqs. (16.5)-(16.7) and (18.1) into account, we have

$$W(E) = (E/2\pi\sigma_V^2) \int_0^{2\pi} \exp\left[-(E^2 \cos^2 \psi + E^2 \sin^2 \psi)/2\sigma_V^2\right] d\psi,$$

i.e.,

$$W(E) = (E/\sigma_V^2) \exp(-E^2/2\sigma_V^2). \tag{18.2}$$

We have thus obtained a Rayleigh distribution. This distribution has been used in a number of papers (see [13, 15, 23, 25, 31], and others) for the analysis of the statistical characteristics of the reverberation envelope.

We now indicate certain parameters of the Rayleigh distribution:

Mean value:

$$\langle E \rangle = \sqrt{\pi/2}\,\sigma_V, \ \langle E^2 \rangle = 2\sigma_V^2;$$

Fluctuation variance:

$$\sigma_E^2 = (4 - \pi)\,\sigma_V^2/2 \approx 0.43\sigma_V^2;$$

Coefficient of variation, which is defined as

$$\gamma_V = \sqrt{\langle E^2 \rangle - \langle E \rangle^2}/\langle E \rangle = \sigma_E/\langle E \rangle, \tag{18.3}$$

and for a Rayleigh distribution turns out to be equal to

$$\gamma_V \approx 0.52.$$

The dimensionless probability density (18.2) has the following form for the normalized quantity E/σ_V:

$$W(E)\sigma_E \approx 0.43 (E/\sigma_E) \exp\left[-0.215 (E/\sigma_E)^2\right]. \tag{18.4}$$

The distribution (18.4) is conveniently compared with the experimental data, from which it is a simple matter to find the mean-square values σ_E of the envelope and to transform to the normalized arguments E/σ_E and the dimensionless values of $W(E)\sigma_E$.

Furthermore, we can determine the probability density W(J) of the reverberation envelope squared $J = E^2$, which is found from the following general relation for the quadratic transformation of stochastic variables:

$$W(J) = (^1/_2 \sqrt{J}) W_E (\sqrt{J}), \; J \geqslant 0, \tag{18.5}$$

where $W_E(E)$ is determined in (18.2). For W(J), therefore, we obtain

$$W(J) = (1/2\sigma_V^2) \exp (- J/2\sigma_V^2),$$

or

$$W(J) = (1/\langle J \rangle) \exp (- J/\langle J \rangle), \tag{18.6}$$

where $<J> = 2\sigma_V^2$ is the mean value of the envelope squared.

The probability $P(J > J_0)$, which characterizes the integral distribution law of the envelope intensity, is found by means of the equation

$$P(J > J_0) = \int_{J_0}^{\infty} W(J) \, dJ$$

and, with allowance for (18.6), is equal to

$$P(J > J_0) = \exp (- J_0/\langle J \rangle). \tag{18.7}$$

We next consider the influence of individual strong scattered signals or coherent scattering on the probability density of the reverberation envelope. If the amplitude of the regular component in the reverberation process is equal to b_0, the probability density $W(E_1)$ of the envelope $E_1(t)$ defined according to (17.5) may be found in correspondence with the general equation

$$W(E_1) = E_1 \int_0^{2\pi} W_V (E_1 \cos \psi_1, \; E_1 \sin \psi_1) \, d\psi_1, \; E_1 \geqslant 0, \tag{18.8}$$

where $W_V(V_{c1}, V_{s1})$ is the joint distribution of the quadrature components of the given process $V_1(t)$.

Under the assumptions made above, taking (17.7) and (18.8) into account, we obtain

$$W(E_1) = (E_1/2\pi\sigma_V^2) \exp [- (E_1^2 + b_0^2)/2\sigma_V^2] \int_0^{2\pi} \exp (E_1 b_0 \cos \psi_1/\sigma_V^2) \, d\psi_1.$$

Integration yields the expression

$$W(E_1) = (E_1/\sigma_V^2) \exp [- (E_1^2 + b_0^2)/2\sigma_V^2] I_0 (E_1 b_0/\sigma_V^2), \tag{18.9}$$

where

$$I_0(X) = (1/2\pi) \int_0^{2\pi} \exp (X \cos \theta) \, d\theta \tag{18.10}$$

is a zero-th order Bessel function of an imaginary argument (modified Bessel function).

We have thus obtained for the reverberation envelope the so-called generalized Rayleigh distribution or Rice distribution.

We note that the distribution (18.9) is valid for the envelope of the sum of normal narrow-band noise and a sinusoidal signal (see [1, 12, 19]). This coincidence is reasonable, since the reverberation distribution was assumed to be normal, while the process b(t) has a constant amplitude.

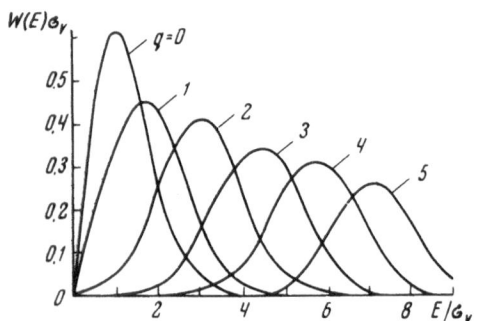

Fig. 17. Distribution of the reverberation envelope in the presence of a regular component of constant amplitude.

A family of curves for the probability density (18.9), constructed for various values of the parameter $q = b_0 / \sqrt{2\sigma_V}$, is shown in Fig. 17.

We see that the distribution (18.9) converges to a normal law for $q \gg 1$. In fact, if we make use of the expansion of the Bessel function $I_0(X)$ for large arguments:

$$I_0(X) = (1/\sqrt{2\pi X}) \exp(X)\,[1$$
$$+ (^1/_8 X) + (^9/_{128} X^2) + \ldots], \quad X \gg 1,$$

then, retaining only the first two terms of this expansion, we write approximately for $W(E_1)$ on the basis of (18.9):

$$W(E_1) \approx (1/\sigma_V)(1 + \sigma_V^2/8E_1 b_0)\,\sqrt{E_1/2\pi b_0}\,\exp\,[-(E_1 - b_0)^2/2\sigma_V^2]. \tag{18.11}$$

For relatively small departures of E_1 from b_0, i.e., for $E_1/b_0 \approx 1$, Eq. (18.11) implies

$$W(E_1) \approx (1/\sqrt{2\pi}\sigma_V)\exp\,[-(E_1 - b_0)^2/2\sigma_V^2]\,(1 + \sigma_V^2/8E_1 b_0), \tag{18.12}$$

where the first term in (18.12) yields a normal distribution with a mean equal to b_0 and variance σ_V^2.

Let us estimate certain parameters of the distribution (18.9) of the reverberation envelope. The k-th order moments $\langle E_1^k \rangle$ about the origin are found from the equation

$$\langle E_1^k \rangle = \int_0^\infty E_1^k W(E_1)\,dE_1. \tag{18.13}$$

Substituting (18.9) into (18.13) and integration leads to the following result:

$$\langle E_1^k \rangle = (2\sigma_V^2)^{k/2}\,\Gamma\,[(k/2) + 1]\,_1F_1\,(-k/2;\ 1;\ -q^2), \tag{18.14}$$

where $\Gamma(X)$ is a gamma function, $_1F_1(X, Y, Z)$ is a confluent hypergeometric function.

Making use of Eq. (18.14), we find the coefficient of variation, defined in (18.3), of the reverberation envelope:

$$\gamma_V = \left[\frac{\Gamma(2)\,_1F_1(-1;\ 1;\ -q^2)}{\Gamma^2(1,5)\,_1F_1^2(-1/2;\ 1;\ -q^2)} - 1\right]^{1/2}. \tag{18.15}$$

If we express the hypergeometric functions in terms of Bessel functions and substitute them into (18.15), also taking into account the values of the gamma function, we obtain for γ_V

$$\gamma_V = \left\{\frac{4(1 + q^2)}{\pi \exp(-q^2)\,[(1 + q^2)\,I_0(q^2/2) + q^2 I_1(q^2/2)]^2} - 1\right\}^{1/2}. \tag{18.16}$$

The dependence of γ_V on the parameter q, as constructed according to Eq. (18.16), is shown in Fig. 18. It follows from an analysis of this dependence that as $q \to \infty$ the value of $\gamma_V \to 0$. This intimates a reduction in the relative fluctuations of the reverberation envelope as the amplitude of the regular scattered signal increases.

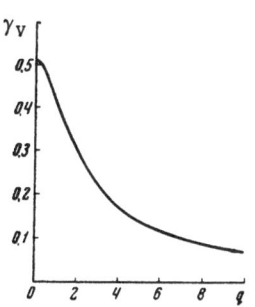

Fig. 18. Dependence of the coefficient of variation on the parameter q.

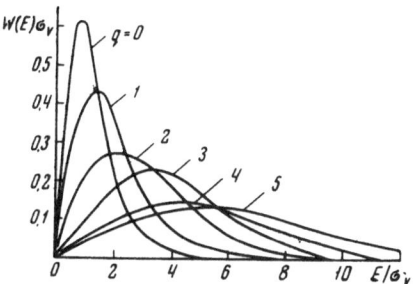

Fig. 19. Distribution of the reverberation envelope in the presence of an individual fluctuating scattered signal.

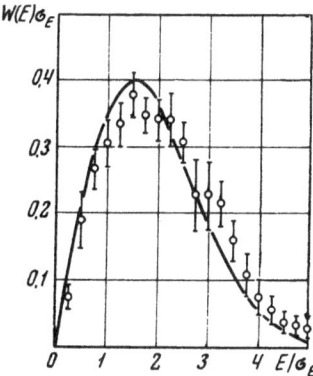

Fig. 20. Summary graph of the measured probability density of the reverberation envelope (circles) and a Rayleigh distribution (curve).

Along with the investigation of regular signals and their influence on the reverberation distribution, it is instructive to analyze the case when in the reverberation process there is a separate fluctuating signal, the distribution of whose instantaneous values is described by a normal law with variance σ_b^2. Assuming statistical independence of the processes V(t) and b(t), we readily obtain the following distribution for the envelope of the total process:

$$W(E_1) = \frac{E_1}{\sigma_V^2(1+q^2)} \exp\left|[-E_1^2/2\sigma_V^2(1+q^2)], \; E_1 \geqslant 0, \right.$$

(18.17)

where

$$q = \sigma_b/\sigma_V.$$

A graph of this distribution for various values of the parameter q is shown in Fig. 19.

The one-dimensional distribution of the envelope and some of its parameters have been investigated under oceanic conditions.

The averaged distributions obtained for various forms of transmitted signals for volume, surface, and bottom reverberation, as well as their aggregate in general, turned out to be close to the theoretical. An example of such a distribution, constructed from the results of processing more than 5000 independent occurrences of various types of reverberation, with five to ten independent readings in each occurrence, is shown in Fig. 20.

In [25, 31], wherein data are presented on the one-dimensional distributions of reverberation envelopes, satisfactory agreement is observed between the experimental and calculated dependences. It is stressed in [31], however, that better agreement with the Rayleigh law is observed for the envelope of volume reverberation, whereas, for surface reverberation, the experimental probability density has a peak shifted to the right of its analytical counterpart and is more aptly described by a gamma distribution. It is indicated in this paper that so far there is not adequate information for the physical substantiation of such an agreement.

It was interesting to investigate the distributions of the reverberation envelope in the transmission of frequency-modulated pulses. The measurements were performed at a frequency of 60 kc for pulses with a duration of 3 msec with the observation of bottom reverberation.

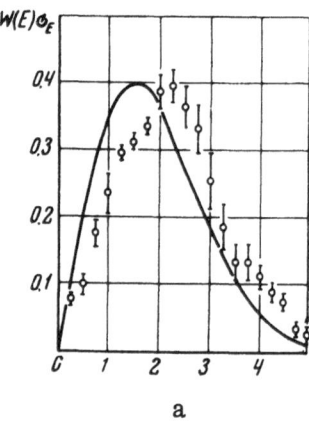

a

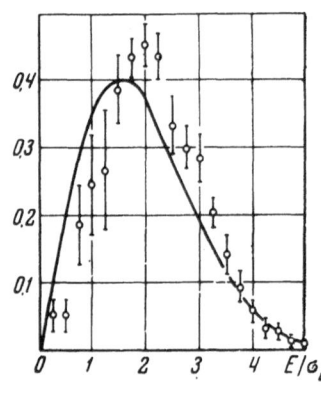

b

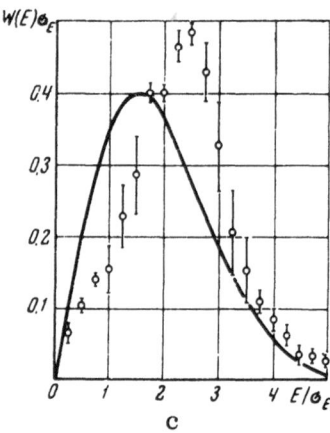

c

Fig. 21. Distributions of the reverberation
envelope for various values of the relative
frequency deviation. a) $\Delta F_M / f_0 = 0$; b) 7%;
c) 17%; $f_0 = 60$ kc; $\delta = 3$ msec. The circles
represent the measured values; the curves
represent the Rayleigh distribution.

Table 18.1. Values of the Coefficient of
Variation for the Transmission of
Frequency-Modulated Pulses

δ, msec	γ_v					
	Frequency deviation					
	0	0.5	1.0	1.5	2.0	2.5
10	0.39	—	0.45	0.41	—	0.38
1	0.47	0.37	—	0.41	0.49	0.43

In the locale of the measurements the
prevalent type of bottom soil was slimey sand.

The measurements were designed for the
purpose of studying the form of the distributions
of reverberation envelopes for various frequency deviations. Experimental data on the probability density of the envelopes are shown in Fig.
21; also shown in the same graph for comparison
are the theoretical relations constructed in accordance with a Rayleigh distribution [see (18.4)].
The experimental data in each series of measurements were processed for occurrences of reverberation, and in each occurrence the stationarized reverberation portion was analyzed, including five to seven independent readings. The
averaged data (circles) are plotted according to
the results of analyzing five series of such measurements.

It is apparent from Fig. 21 that as the
relative frequency deviation increases from
zero to 17%, the nature of the distribution curves
does not alter, and in every case a typical shift
of the measured peak probability density to the
right of the analytical value is observed. This
shift may clearly be attributed to the presence
of individual large-scale scattering centers in
the bottom, for example, rocks, causing the ap-
pearance of a regular component in the reverberation process.

Another series of measurements was carried out for surface reverberation, with a view
toward clarifying the dependence of the coefficient of variation of the envelope on the frequency deviation.

The general conclusion one draws from the results of the measurements is that the coefficient of variation turns out to be practically independent of the frequency deviation.

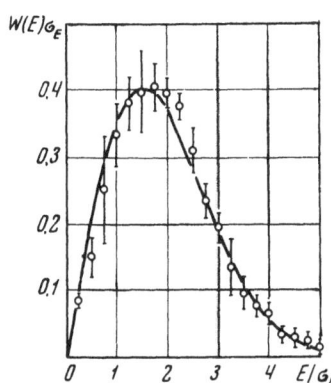

Fig. 22. Distribution of the reverberation envelope for the transmission of a noisy signal. δ = 3 msec, ΔF_N = 4 kc, f_0 = 36 kc. The circles represent the measured values; the curve represents a Rayleigh distribution.

The results of the experimental processing of a surface reverberation envelope produced by the transmission of frequency-modulated signals with a rectangular envelope are presented in Table 18.1 as an example. The coefficient of variation was computed for two values indicated in the table for the duration of the transmitted pulses.

The distributions of the reverberation envelope were also investigated for the transmission of noisy pulses. The envelope probability density obtained for such pulses is shown in Fig. 22. Comparison of the experimental data with a Rayleigh distribution shows that the observed distribution is close to the theoretical, which is quite reasonable to expect, even when the reverberation process is generated by a small number of elementary scattered signals, because each perturbation is distributed according to a normal law in this case.

The experimental intensity distribution of the envelope and the curve corresponding to the law (18.7) are shown in Fig. 23.

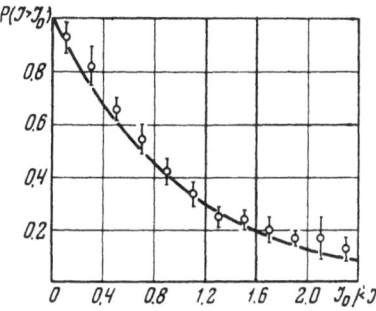

Fig. 23. Comparison of the measured integral, intensity distribution of a reverberation envelope (circles) with the theoretical law (18.7).

§ 19.   Distributions in the Summation of Reverberation Processes

For the solution of a number of problems in the processing of underwater acoustical information, it is required to know the probability distribution of a summation of reverberation processes or their envelopes.

Let the instantaneous values of N reverberation processes be summed, so that the resultant process c(t) is represented in the form

$$c\,(t) = \sum_{i=1}^{N} V_i\,(t). \tag{19.1}$$

For normal distributions of the components, a normal distribution is also valid for the probability density W(c), with a zero mean value and variance σ_c^2 equal to

$$\sigma_c^2 = \sum_{i=1}^{N} \sigma_{V_i}^2 + \sum_{\substack{i \neq j \\ i,\,j=1}}^{N} \sigma_{V_i} \sigma_{V_j} R_{ij}, \tag{19.2}$$

where $\sigma_{V_i}^2$ is the variances of the i-th process $V_i(t)$, R_{ij} is the cross-correlation coefficient of the components $V_i(t)$ and $V_j(t)$:

$$R_{ij} = \langle V_i(t) V_j(t) \rangle / \sigma_{V_i} \sigma_{V_j}.$$ (19.3)

In particular, for uncorrelated components with the same variance, equal to σ_V, we have

$$\sigma_c^2 = N \sigma_V^2.$$ (19.4)

We next consider the probability distributions of a stochastic process that is obtained as the result of summation of the envelopes of several reverberation signals:

$$c(t) = \sum_{i=1}^{N} E_i(t)$$ (19.5)

and their squares:

$$c(t) = \sum_{i=1}^{N} E_i^2(t) = \sum_{i=1}^{N} J_i(t).$$ (19.6)

We will assume that the components in the sums (19.5) and (19.6) are statistically independent. This case might occur, for example, in the observation of several uncorrelated reverberation processes at different points in space or in the investigation of a group of pulses with different carrier frequencies, as well as in a number of other cases.

We begin our discussion with an analysis of the probability density of the sum (19.5).

If we assume that the variances of the processes $E_i(t)$ are identical, we have for the probability density $W(E_i)$ of any i-th component, taking Eq. (18.2) into account:

$$W(E_i) = (E_i / \sigma_V^2) \exp(- E_i^2 / 2\sigma_V^2).$$ (19.7)

The distribution W(c) may be found either by the method of characteristic functions or by representation of the sought-after probability density in the form of a series (see Sec. 8). We will use the latter method here. The essential fact is that the characteristic function $\theta(\eta)$ of the distribution (19.7) is expressed by the relation

$$\theta(\eta) = 1 + j\sigma_V \eta \sqrt{\pi/2} \, [1 + \Phi(j\sigma_V \eta / \sqrt{2})] \exp(- \sigma_V^2 \eta^2 / 2),$$ (19.8)

where $\Phi(X)$ is the probability integral. If we attempt to find the probability density of the sum (19.5), applying the general expression

$$W(C) = (1/2\pi) \int_{-\infty}^{\infty} \theta^N(\eta) \exp(- jC\eta) \, d\eta,$$

it proves unfeasible upon substitution of (19.8) therein, for N > 1, to compute exactly the integrals by which W(C) is determined. It is possible, of course, to use series expansions of $\theta(\eta)$ of $\exp(-jC\eta)$, but this leads essentially to the analytical method involved in the representation of the probability density as an Edgeworth series.

In this connection we use the relation (8.13) and write the unknown probability density in the form

$$W(C) \approx (1/\sqrt{2\pi}\sigma_c) \exp[-(C - \langle C \rangle)^2 / 2\sigma_c^2] \{ 1 + (\gamma_a/3!) H_3 [(C - \langle C \rangle)/\sigma_c] +$$
$$+ (\gamma_e/4!) H_4 [(C - \langle C \rangle)/\sigma_c] + (10\gamma_a^2/6!) H_6 [(C - \langle C \rangle)/\sigma_c] \},$$ (19.9)

where γ_a and γ_e are the coefficients of skewness (asymmetry) and excess of the distribution W(C); $<C>$ and σ_c are the mean and mean-square values of the process c(t); the $H_k(X)$ are Hermite polynomials.

We now introduce the coefficients γ_{a1} and γ_{e1} of the distribution (19.7) of the components and allow for the fact that

$$\left.\begin{array}{ll} \langle C \rangle = N \langle E \rangle, & \gamma_a = \gamma_{a1}/\sqrt{N}, \\ \sigma_c^2 = N\sigma_E^2, & \gamma_e = \gamma_{e1}/N. \end{array}\right\} \tag{19.10}$$

Then, with due regard for the values of γ_{a1} and γ_{e1} for the initial Rayleigh distribution,

$$\gamma_{a1} = 0.67, \qquad \gamma_{e1} = 0.25,$$

and making use of the expressions for the third-, fourth-, and sixth-order Hermite polynomials (8.5), we obtain for the probability density W(C),

$$W(C) \approx (1/\sqrt{2\pi N}\sigma_E) \exp\left[-\frac{(C-N\langle E\rangle)^2}{2N\sigma_E^2}\right]\left[1 - 0.053/N\right.$$

$$- (0.315/\sqrt{N})\frac{C-N\langle E\rangle}{\sqrt{N}\sigma_E} + (0.188/N)\left(\frac{C-N\langle E\rangle}{\sqrt{N}\sigma_E}\right)^2$$

$$+ (0.105/\sqrt{N})\left(\frac{C-N\langle E\rangle}{\sqrt{N}\sigma_E}\right)^3 - (0.073/N)\left(\frac{C-N\langle E\rangle}{\sqrt{N}\sigma_E}\right)^4 + (0.006/N)\left(\frac{C-N\langle E\rangle}{\sqrt{N}\sigma_E}\right)^6\right]. \tag{19.11}$$

It is readily seen from (19.11) that as $N \to \infty$, the distribution W(C) converges to a normal distribution with mean value $<C>$ and variance σ_c^2 determined by the relations (19.10).

A family of curves constructed for the investigated probability density for various numbers of components is shown in Fig. 24, where curve 1 is drawn for N = 1 in correspondence with the exact formula for the Rayleigh distribution, the rest are drawn according to the approximate equation (19.11).

We proceed next with an analysis of the distribution of the sum of the squares of the envelopes of reverberation processes. The probability density for the sum (19.6) may be found by the method of characteristic functions. On the basis of (18.6) we have for the characteristic function of any term

$$\Theta(\eta) = (1/\langle J\rangle) \int_0^\infty \exp(jJ\eta - J/\langle J\rangle)\,dJ,$$

whence

$$\Theta(\eta) = (1 - j\langle J\rangle\eta)^{-1}. \tag{19.12}$$

The characteristic function $\Theta_C(\eta)$, based on Eq. (19.12) for the total process, is equal to

$$\Theta_C(\eta) = (1 - j\langle J\rangle\eta)^{-N},$$

which means that the unknown probability density may be written in the form

$$W(C) = (1/2\pi) \int_{-\infty}^\infty (1 - j\langle J\rangle\eta)^{-N} \exp(-jC\eta)\,d\eta.$$

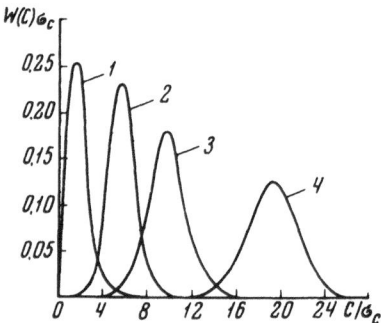

Fig. 24. Probability density of a sum of the envelopes of reverberation signals. 1) Rayleigh distribution (N = 1); 2) N = 3; 3) N = 5; 4) N = 10.

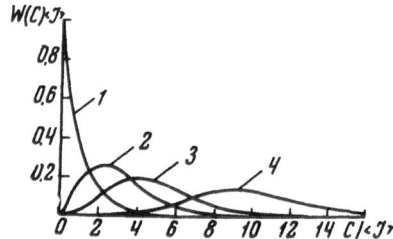

Fig. 25. Probability density of the sum of the squares of reverberation signal envelopes (gamma distribution). 1) N = 1; 2) N = 3; 3) N = 5; 4) N = 10.

We now invoke the tabulated formula [9]:

$$\int_{-\infty}^{\infty} (\beta - jX)^{-N} \exp(-jpX)\,dX = [2\pi p^{N-1}/\Gamma(N)]\exp(-\beta p),$$

where $\Gamma(N)$ is a gamma function.

Then for the probability density $W(C)$ we obtain

$$W(C) = [C^{N-1}/\Gamma(N)\langle J\rangle^N]\exp(-C/\langle J\rangle); \quad C \geqslant 0. \tag{19.13}$$

This probability density is known as the gamma distribution.

Some of its parameters, viz., the mean value, variance, coefficients of skewness and excess, are equal to, respectively,

$$\left.\begin{array}{l} \langle c\rangle = N\langle J\rangle, \quad \gamma_a = 2/\sqrt{N}, \\ \sigma_c^2 = N\langle J\rangle^2, \quad \gamma_e = 6/N. \end{array}\right\} \tag{19.14}$$

Graphs of the probability density (19.13) are shown in Fig. 25 for various numbers of components.

For large N, the investigated distribution (19.13) converges to a normal law. This is natural, since we are dealing with the probability density of a normalized sum of statistically independent stochastic variables.

In [31, 49], wherein the distribution of scattered signals obtained in the addition of sine waves with different carrier frequencies and random phases is analyzed, the authors also arrived at a gamma distribution for the intensity of the resultant sound field. In [49], in particular, the coefficient of variation of this distribution was computed, and it was shown that with an increase in the number of terms this coefficient would tend to zero, while the distribution itself would transform to a delta function in the case of a finite variance for the total process.

The results of the indicated papers could prove significant for the investigation of the properties of reverberation generated by scattering at moving inhomogeneities or in multifrequency transmission.

CHAPTER IV

CORRELATION ANALYSIS OF REVERBERATION

§20. Preliminary Remarks

In the preceding chapter we investigated the probability distributions of the instantaneous reverberation values, its phase and envelope, with primary attention centered on the one-dimensional distributions. The next step in studying the statistical properties of reverberation is a correlation analysis.

The correlation characteristics of reverberation are of particular interest in connection with the fact that the distributions of their instantaneous values are described either by a normal law or by a law similar to it (see Chapter III). This means that the results of correlation analysis may be used in a number of cases in principle as the basis for finding the multidimensional distributions, which completely determine the properties of reverberation signals in the statistical sense. On the other hand, various linear and nonlinear transformations of reverberation signals in connection with the investigation of the properties of the envelope, phase, and functions of these lead to processes arising out of normal ones; such processes are described by distribution functions derivable from a normal law. The multidimensional normal distribution is completely determined by the correlation functions (correlation matrix); hence, the comprehensive description of processes deriving from normal processes is also possible when the correlation characteristics of the primitive process are known.

In the present chapter we analyze the autocorrelation functions and cross-correlation functions of reverberation with varying degrees of idealization of the scattering process.

We first examine the autocorrelation characteristics of reverberation on the assumption that the elementary scattered signals reproduce the shape of the transmitted pulses. This analysis is based on the theorems of superposition of stochastic perturbations, as formulated in Secs. 5 and 6. The remainder of the present chapter is devoted to certain additional investigations of reverberation correlation with a view toward elucidating the spatial and frequency-time statistical relations in reverberation signals, as well as an analysis of the influence of motion of the scatterers and displacement of the acoustic receiving and transmitting arrays; this analysis is based on the results of Sec. 7.

§21. General Relations for the Correlation Function of the Instantaneous Values

In accordance with the two-dimensional theorem of superposition of stochastic perturbations (see Sec. 6, Corollary 2) the autocorrelation function $B_r(\tau)$ of reverberation may be written in the following form when the form of all the elementary scattered signals is the same:

$$B_{\mathrm{r}}(\tau) = \langle n_1 \rangle \langle a^2 \rangle \int\limits_{-\infty}^{\infty} s(t)\, s(t + \tau)\, dt. \qquad (21.1)$$

The correlation function $R_{\mathrm{r}}(\tau)$ is defined, therefore, by the relation

$$R_{\mathrm{r}}(\tau) = \frac{\displaystyle\int\limits_{-\infty}^{\infty} s(t)\, s(t+\tau)\, dt}{\displaystyle\int\limits_{-\infty}^{\infty} s^2(t)\, dt}. \qquad (21.2)$$

We recall that the stationary component of reverberation is under consideration here, i.e., the stochastic function V(t), which is obtained after stationarization of the reverberation process according to its variance. It may be assumed, therefore, that the product $\langle n_1 \rangle \langle a^2 \rangle$ does not depend on the time over a sufficiently long time interval $(-T/2,\ T/2)$, considerably larger than the effective duration of the transmitted signal $(T \gg \delta_{\mathrm{ef}})$.

In general form the function s(t) is written as follows for determinate quasi-harmonic transmitted signals (see Sec. 10):

$$s(t) = s_0(t) \cos[\omega_0 t + \Phi(t)], \qquad (21.3)$$

where $s_0(t)$ is the signal envelope, ω_0 is the central frequency of its spectrum, $\Phi(t)$ is a function governing the frequency modulation. Then we obtain for the correlation coefficient, taking (21.1) and (21.3) into account,

$$R_{\mathrm{r}}(\tau) = (1/\delta_{\mathrm{ef}}) \left\{ \cos \omega_0 \tau \int\limits_{-\infty}^{\infty} s_0(t)\, s_0(t + \tau) \cos[\Phi(t + \tau) - \Phi(t)]\, dt - \right.$$

$$\left. - \sin \omega_0 \tau \int\limits_{-\infty}^{\infty} s_0(t)\, s_0(t + \tau) \sin[\Phi(t + \tau) - \Phi(t)]\, dt \right\}. \qquad (21.4)$$

For quasi-harmonic signals of the given type, clearly, the following inequality holds:

$$\int\limits_{\infty}^{\infty} s_0(t)\, s_0(t + \tau) \cos\,]\Phi(t + \tau) - \Phi(t)]\, dt \gg \int\limits_{-\infty}^{\infty} s_0(t)\, s_0(t + \tau) \sin[\Phi(t + \tau) - \Phi(t)]\, dt,$$

and, consequently, the second integral in (21.4) may be neglected. We thus have for the reverberation correlation coefficient,

$$R_{\mathrm{r}}(\tau) \approx (1/\delta_{\mathrm{ef}}) \left\{ \int\limits_{-\infty}^{\infty} s_0(t)\, s_0(t + \tau) \cos[\Phi(t + \tau) - \Phi(t)]\, dt \right\} \cos \omega_0 \tau. \qquad (21.5)$$

It is convenient also to make use of the following notation:

$$R_{\mathrm{r}}(\tau) = r_{\mathrm{r}}(\tau) \cos \omega_0 \tau, \qquad (21.6)$$

where the function

$$r_{\mathrm{r}}(\tau) \approx (1/\delta_{\mathrm{ef}}) \int\limits_{-\infty}^{\infty} s_0(t)\, s_0(t + \tau) \cos[\Phi(t + \tau) - \Phi(t)]\, dt \qquad (21.7)$$

could be called the <u>correlation coefficient envelope</u> of the instantaneous reverberation values. For signals of the quasi-harmonic type, this function varies considerably more slowly than $\cos \omega_0 \tau$.

We further consider the relationships for the correlation coefficient when the radiation consists of pulses of finite length:

$$s(t) = 0, \quad |t| \leqslant t_1/2. \tag{21.8}$$

In this case, $r_\mathrm{r}(\tau)$ may be written in the following form on the basis of (21.7) and (21.8):

$$r_\mathrm{r}(\tau) \approx (2/\delta_\mathrm{ef}) \int_0^{(t_1-|\tau|)/2} s_0(t - \tau/2) s_0(t + \tau/2) \cos \left[\Phi(t + \tau/2) - \Phi(t - \tau/2) \right] dt. \tag{21.9}$$

The relations just obtained enable one to calculate the autocorrelation characteristics of reverberation for the investigation of signals of any shape of the determinate type.

For the correlation analysis of stochastic processes certain integral parameters are sometimes used to characterize the effective correlation time, i.e., in the given case the width of the curve $R_\mathrm{r}(\tau)$. One such parameter is the <u>correlation interval</u>, which is suitably defined for quasi-harmonic processes in terms of the correlation coefficient envelope $r_\mathrm{r}(\tau)$.

The reverberation correlation time interval τ_r may in particular be defined as follows:

$$\tau_\mathrm{r} = \int_{-\infty}^{\infty} |r_\mathrm{r}(\tau)| \, d\tau. \tag{21.10}$$

It characterizes the time scale of the envelope or phase fluctuations, i.e., the period in which the reverberation remains coherent.

§ 22. Transmission of Pulses with a Sinusoidal Carrier and Various Envelope Profiles

Let the transmitted signal represent a sinusoidal vibration with an arbitrary envelope profile $s_0(t)$. Then, with regard for the fact that $\Phi(t) = 0$, we obtain from Eq. (21.5)

$$R_\mathrm{r}(\tau) \approx (1/\delta_\mathrm{ef}) \left[\int_{-\infty}^{\infty} s_0(t) s_0(t + \tau) \, dt \right] \cos \omega_0 \tau. \tag{22.1}$$

The relation (22.1) shows that the correlation of reverberation depends only on the form of the transmitted signal envelope.

The correlation coefficient envelope in this case is defined as follows:

$$r_\mathrm{r}(\tau) \approx (1/\delta_\mathrm{ef}) \int_{-\infty}^{\infty} s_0(t) s_0(t + \tau) \, dt. \tag{22.2}$$

If the duration of the transmitted signals is finite and equal to t_1, we have on the basis of Eqs. (21.9) and (22.2),

Table 22.1. Autocorrelation Coefficient Envelopes of Correlation Interval of Reverberation for Certain Types of Signals

No.	Type of signal envelope		$r_{\mathrm{r}}(\tau)$	τ_{r}
1	Rectangular pulse	$1, |t| \leqslant \delta/2$	$1 - |\tau|/\delta,\ |\tau| \leqslant \delta$	$\delta/2$
2	Bell-shaped pulse	$\exp(-t^2/t_0^2)$	$\exp[-(\pi/4)(\tau/\delta_{\mathrm{ef}})^2]$	δ_{ef}
		$\exp(-t^2/t_0^2),$ $|t| \leqslant t_1/2$	$\sqrt{\pi/2}\,(t_0/\delta_{\mathrm{ef}})\,\Phi\,\dfrac{(t_1-|\tau|)}{\sqrt{2}\,t_0}\exp(-\tau^2/2t_0),$ $|\tau| \leqslant t_1$	—
3	Cosine to the p-th power	$\cos^p(\pi t/t_1),$ $|t| \leqslant t_1/2$	$1/\delta_{\mathrm{ef}}\Big\{[(t_1-|\tau|)/2]\cos(\pi|\tau|/t_1) + (t_1/2\pi)\sin(\pi|\tau|/t_1)\Big\}\quad |\tau| \leqslant t_1,\ p=1$	$(4/\pi^2)\,t_1 \approx 0.39t$
			$1/\delta_{\mathrm{ef}}\{[(t_1-|\tau|)/4] + [(t_1-|\tau|)/8]\times \cos(2\pi|\tau|/t_1) + (t_1/4\pi)\sin(2\pi|\tau|/t_1) - t_1/16\pi\sin(2\pi|\tau|/t_1)\},\ |\tau| \leqslant t_1,\ p=2$	$(1/3)\,t_1 \approx 0.33t_1$
4	Exponential pulse	$\exp(-t/t_0),\ t \geqslant 0$ $\exp(-t/t_0),$ $t \geqslant 0,\ t < t_1$	$\exp(-|\tau|/t_0)$ $\dfrac{\exp(-|\tau|/t_0)\{1 - \exp[-(2t_1/t_0)(1-|\tau|/t_1)]\}}{1 - \exp(-2t_1/t_0)}$	$2\delta_{\mathrm{ef}}$ —
5	Amplitude-modulated rectangular pulse	$\dfrac{1 + m\cos\Omega_1 t}{1 + m},$ $|t| \leqslant t_1/2,\ \Omega_1\delta \gg 1$	$(t_1 - |\tau|/\delta_{\mathrm{ef}})\dfrac{1 + (m^2/2)\cos\Omega_1\tau}{(1+m)^2},$ $|\tau| \leqslant t_1$	—

$$r_{\mathrm{r}}(\tau) \approx (2/\delta_{\mathrm{ef}}) \int_0^{(t_1-|\tau|)/2} s_0(t-\tau/2)\,s_0(t+\tau/2)\,dt. \qquad (22.3)$$

The correlation characteristics of reverberation corresponding to transmitted signals with the most important envelope profiles from the practical point of view are shown in Table 22.1.

We recall that the values of the effective pulse durations were given in Table 11.1.

It is interesting to note that for $t_1 \gg t_0$ the correlation characteristics of reverberation depend mainly on the form of the envelope $s_0(t)$ for different signal envelopes. If $t_1 \ll t_0$, however, the form of the envelope degenerates into a rectangular profile in every case:

$$r_{\mathrm{r}}(\tau) \approx (1 - |\tau|/\delta_{\mathrm{ef}}),\quad |\tau| \leqslant \delta_{\mathrm{ef}}, \qquad (22.4)$$

i.e., the reverberation correlation proves to be the same as for the transmission of rectangular pulses.

Let us consider the particular features of the reverberation correlation in the transmission of amplitude-modulated pulses.

It is apparent from row 5 of Table 22.1 that the modulation index of the reverberation correlation coefficient differs from the modulation index in the transmitted signal. If we define the modulation index of the function $r_{\mathrm{r}}(\tau)$ as

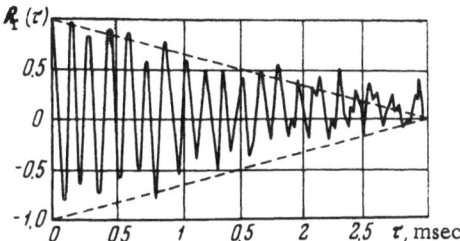

Fig. 26. Measured reverberation autocorrelation coefficient for the transmission of rectangular pulses with a duration of 3 msec. The dashed lines represent the calculated autocorrelation coefficient envelope.

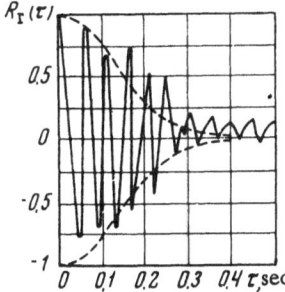

Fig. 27. Measured reverberation autocorrelation coefficient for the transmission of pulses with a bell-shaped envelope and effective duration of 150 msec. The dashed curves represent the calculated autocorrelation coefficient envelope.

$$m_k = \frac{R_{\mathrm{r\ max}}(\tau) - R_{\mathrm{r\ min}}(\tau)}{R_{\mathrm{r\ max}}(\tau) + R_{\mathrm{r\ min}}(\tau)} , \quad |\tau| \leqslant t_1,$$

it turns out that

$$m_k = m^2/2.$$

This is reasonable, since the correlation characteristics are referred to the energy parameters.

The experimental investigation of the autocorrelation characteristics of reverberation in the transmission of pulses with various envelope profiles exhibited good agreement between the measured data and the calculated dependences. For example, Figs. 26 and 27 show the results of measurements of the reverberation autocorrelation coefficients in the case of transmitted pulses with rectangular and bell-shaped envelopes. In the experiments, reverberation was generated by scattering at inhomogeneities of the ocean bottom and surface layer. Individual reverberation events were subjected to processing, and time averaging was carried out within stationarized reverberation intervals, including 40 to 50 independent readings. The analyzer in this case was an analog correlator operating at sonic frequencies, so that the frequency of the investigated process was first converted into the low-frequency range. The oscillation period of the measured correlation coefficient therefore did not correspond to the period of the transmitted-signal carrier frequency.

Let us examine further the correlation of reverberation in the transmission of a pulse train.

If each pulse in the train represents a harmonic oscillation with envelope $s_{01}(t)$, and their repetition period is constant and equal to T_0, a relation analogous to (10.19) applies to the envelope of such a signal, i.e.,

$$s_0(t) = \sum_{k=0}^{N-1} s_{01} [t - kT_0 + (N-1)T_0/2], \tag{22.6}$$

where N is the number of pulses in the train. Equation (22.6) presupposes that the individual pulses of the train do not overlap and are quasi-harmonic functions.

The reverberation correlation coefficient envelope is determined on the basis of (22.2) and (22.6) by the expression

$$r_{\mathrm{r}}(\tau) \approx (1/\delta_{\mathrm{ef}}) \int_{-\infty}^{\infty} \sum_{k=0}^{N-1} \sum_{m=0}^{N-1} s_{01} [t - kT_0 + (N-1)T_0/2] \, s_{01} [t + \tau + mT_0 + (N-1)T_0/2] \, dt, \tag{22.7}$$

where δ_{ef} is the effective transmitted signal duration, calculated according to the equation

$$\delta_{ef} = \int_{-\infty}^{\infty} \Big\{ \sum_{k=0}^{N-1} s_{01} [t - kT_0 + (N-1) T_0/2] \Big\}^2 dt. \tag{22.8}$$

In order to simplify Eq.(22.7), we introduce certain postulates regarding the structure of the pulse train. We assume that the individual pulses in the train have an effective duration

$$\delta_{ef_1} = \int s_{01}^2 (t) \, dt, \tag{22.9}$$

which is less than half their period of separation, i.e.,

$$\delta_{ef_1} < T_0/2.$$

Then, for the total train, we have the approximate relation

$$\delta_{ef} \approx N \delta_{ef_1}. \tag{22.10}$$

If we also presume that the number of pulses in the train is large, i.e., $N \gg 1$, then from (22.7), taking (22.10) into account, we arrive at the approximate expression

$$r_{\tau} (\tau) \approx (1/\delta_{ef_1}) (1 - |\tau|/NT_0) \sum_{k=0}^{N-1} \int_{-\infty}^{\infty} s_{01} [t - (N-1) T_0/2] \, s_{01} [t + \tau - (N-1) T_0/2] \, dt, \tag{22.11}$$

$$|\tau| \leqslant NT_0.$$

Since the correlation coefficient of the k-th single pulse is defined as

$$r_{\tau 1} (\tau - kT_0) \approx (1/\delta_{ef})^2 \int_{-\infty}^{\infty} s_{01} [t - (N-1) T_0/2] \, s_{01} [t + \tau - (N-1) T_0/2] \, dt,$$

Eq. (22.11) may be written in the more compact form

$$r_{\tau} (\tau) \approx (1 - |\tau|/NT_0) \sum_{k=0}^{N-1} r_{\tau_1}(\tau - kT_0), \quad |\tau| \leqslant NT_0. \tag{22.12}$$

We note that Eqs.(22.11) and (22.12) are approximate, not only for the reasons mentioned above, but also because, according to them, the correlation decreases with increasing τ according to the linear law $1 - \tau/NT_0$, whereas, in reality, this factor is more complex, being determined in the first approximation by the set of relations

$$\left.\begin{array}{ll}
1 & \text{for} \quad 0 \leqslant \tau < (T_0 - \delta_{ef})/2, \\
1 - 1/N & \text{for} \quad (T_0 - \delta_{ef})/2 \leqslant \tau < [(T_0 - \delta_{ef})/2] + T_0 \\
\cdots\cdots\cdots\cdots\cdots\cdots\cdots\cdots\cdots\cdots\cdots \\
1 - k/N & \text{for} \quad [(T_0 - \delta_{ef})/2] + (k-1)T_0 \leqslant \tau < [(T_0 - \delta_{ef})/2] + kT_0, \\
\cdots\cdots\cdots\cdots\cdots\cdots\cdots\cdots\cdots\cdots\cdots \\
1 - (N-1)/N & \text{for} \quad [(T_0 - \delta_{ef})/2] + (N-2) T_0 \leqslant \tau < \infty.
\end{array}\right\} \tag{22.13}$$

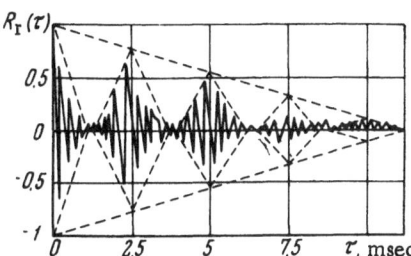

Fig. 28. Measured reverberation auto-correlation coefficient for the transmission of a rectangular-pulse train. N = 5, δ_1 = 1 msec, T_0 = 2.5 msec. The dashed lines represent the calculated autocorrelation coefficient envelope.

If the pulses in the train are dissimilar or their periods of separation are not equal, the equations turn out to be more complex, but the reverberation correlation coefficient is calculated in a manner similar to the above. We see that the general relations presented here may also be used to calculate $r_r(\tau)$ if the amplitudes of the pulses in the trains are not the same.

As an example, we consider the case of transmission of a train of rectangular pulses. The correlation coefficient envelope turns out on the basis of the data in Table 22.1 (row 1) and Eq. (22.12) to be equal to

$$r_r(\tau) \approx (1 - |\tau|/NT_0) \sum_{k=0}^{N-1} (1 - |\tau - kT_0|/\delta_1), \quad |\tau - kT_0| \leqslant \delta_1. \text{ (22.14)} \tag{22.14}$$

Experimental data obtained on the reverberation correlation coefficient for combined surface and bottom reverberation in the transmission of a train of rectangular pulses are shown in Fig. 28. The measurement procedure was similar to that described above in the present section.

As apparent from the graph, the relation (22.14) concurs satisfactorily with the experimental data.

§ 23. Transmission of Wide-Band Signals

In the observation of reverberation produced by the transmission of signals with a sinusoidal carrier, its correlation characteristics, as shown in Sec. 22, are uniquely determined by the form of the signal envelope.

If the transmitted signals are frequency-modulated vibrations or noise pulses, the correlation characteristics of the reverberation will no longer depend on the form of the envelope under certain conditions.

We first investigate the correlation of reverberation for frequency-modulated signals.

The general relation (21.7) enables us to calculate the correlation coefficient envelope for any form of the function $\Phi(t)$ characterizing the law of frequency modulation of the signal. However, in practice it is only possible to carry through the calculations for certain types of modulation functions. We will therefore consider the simplest case, assuming the modulation law to be linear.

Making use of Eqs. (10.12) and (21.7), we obtain for $r_r(\tau)$

$$r_r(\tau) \approx (1/\delta_{ef}) \int_{-\infty}^{\infty} s_0(t)\, s_0(t + \tau) \cos\left[(\Delta\omega_M/2t_M)(2t\tau + \tau^2)\right] dt, \tag{23.1}$$

where $\Delta\omega_M$ is the frequency deviation, t_M is a parameter characterizing the rate of change of frequency.

Let us examine the case when the signal envelope has the rectangular shape (10.7). Then, letting $t_M = \delta$, we go from (23.1) to the following integral representation:

$$r_{\mathrm{r}}(\tau) \approx (2/\delta) \int_0^{(\delta - |\tau|/2)} \cos\left[(\Delta\omega_M\tau/\delta)\, t + \Delta\omega_M\tau^2/2\delta\right] dt, \quad |\tau| \leqslant \delta.$$

Hence we obtain the approximate relation

$$r_{\mathrm{r}}(\tau) \approx \frac{\sin\left[(\Delta\omega_M\tau/2)(1 - |\tau|/\delta)\right]}{\Delta\omega_M\tau/2}. \tag{23.2}$$

If the frequency deviation is small, and if the inequality

$$\Delta F_M\delta \ll 1, \qquad \Delta\omega_M = 2\pi\Delta F_M, \tag{23.3}$$

is satisfied, then

$$\sin\left[(\Delta\omega_M\tau/2)(1 - |\tau|/\delta)\right] \approx (\Delta\omega_M\tau/2)(1 - |\tau|/\delta),$$

and we arrive at the previously derived expression for the reverberation correlation coefficient envelope corresponding to transmission of a pulse with a rectangular envelope and sinusoidal carrier (Table 22.1, row 1).

On the other hand, if the frequency deviation is considerable, so that the following inequality holds:

$$\Delta F_M\delta \gg 1, \tag{23.4}$$

it follows from (23.2) that

$$r_{\mathrm{r}}(\tau) \approx \frac{\sin(\Delta\omega_M\tau/2)}{\Delta\omega_M\tau/2}, \tag{23.5}$$

i.e., the reverberation correlation does not depend on the pulse duration and is determined solely by the frequency deviation.

The indicated characteristics of the reverberation correlation in the transmission of frequency-modulated signals are illustrated by the graph in Fig. 29, which shows a family of curves for $r_{\mathrm{r}}(\tau)$ constructed in correspondence with Eq. (23.2) for various values of the parameter $\Delta F_M\delta$.

If the form of the signal envelope is not rectangular, the expressions for the reverberation correlation coefficients may be found by the described analytical procedure.

Thus, in particular, for a bell-shaped envelope we obtain from (10.8) and (23.1) for $r_{\mathrm{r}}(\tau)$

$$r_{\mathrm{r}}(\tau) \approx \exp\left\{-(\pi/4)(\tau/\delta_{\mathrm{ef}})^2\left[1 + (1/\pi^2)(\Delta\omega_M\delta_{\mathrm{ef}})^2\right]\right\}. \tag{23.6}$$

Hence, it follows that for a small frequency deviation, when an inequality similar to (23.3) applies, the correlation coefficient envelope is approximately equal to

$$r_{\mathrm{r}}(\tau) \approx \exp\left[-(\pi/4)(\tau/\delta_{\mathrm{ef}})^2\right]. \tag{23.7}$$

This expression coincides with the one derived earlier (see Table 22.1, row 2).

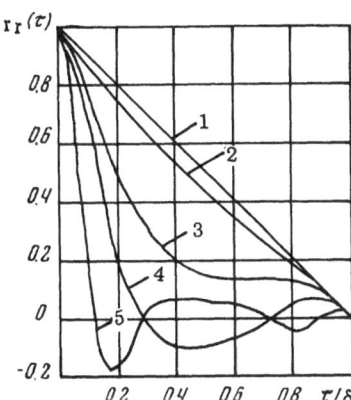

Fig. 29. Calculated reverberation auto-correlation coefficient envelopes in the case of transmission of frequency-modulated signals. 1) $\Delta F_M \delta = 0$; 2) $\Delta F_M \delta = 1$; 3) $\Delta F_M \delta = 3$; 4) $\Delta F_M \delta = 5$; 5) $\Delta F_M \delta = 10$.

On the other hand, if the frequency deviation is large, upon fulfillment of a condition similar to (23.4) we find from (23.6)

$$r_r(\tau) \approx \exp\left[-(1/4\pi)(\Delta\omega_M\tau)^2\right]. \qquad (23.8)$$

Consequently, in the present instance, as in the transmission of pulses with a rectangular envelope, there is no dependence of the correlation coefficient envelope on the signal duration for large frequency deviation. Here, however, the form of the correlation of the curve is determined not only by the law of frequency variation, but by the form of the transmitted-signal envelope as well, which is apparent from a comparison of the relations (23.5) and (23.8).

An estimate of the reverberation correlation interval may be obtained using Eq. (21.10). Performing, for example, the substitution of Eq. (23.6) into (21.10) and carrying out the integration, we find

$$\tau_r \approx \delta_{ef}\left[1 - (1/\pi^2)(\Delta\omega_M\delta_{ef})^2\right]^{-1/2}. \qquad (23.9)$$

An analysis of the expression (23.9) shows that, beginning with $\Delta\omega_M\delta_{ef} > 5$, the correlation interval becomes practically independent of the signal duration and is determined by the frequency modulation effect. For example, for very large frequency deviations,

$$\tau_r \approx \pi/\Delta\omega_M, \qquad (23.10)$$

i.e., the correlation interval is inversely proportional to the value of $\Delta\omega_M$.

We now look further into the correlation characteristics of reverberation in the transmission of noisy signals of the type (10.14).

Taking Eq. (6.28) into consideration, as well as the condition of quasi-harmonicity characterizing the noise carrier of the process x(t), according to which

$$R_x(\tau) = \frac{\langle x(t)\,x(t+\tau)\rangle}{\langle x^2(t)\rangle} = r_x(\tau)\cos\omega_0\tau, \qquad (23.11)$$

we obtain for the reverberation correlation coefficient

$$R_r(\tau) \approx \left[(1/\delta_{ef})\int_{-\infty}^{\infty} s_0(t)\,s_0(t+\tau)\,dt\right]r_x(\tau)\cos\omega_0\tau, \qquad (23.12)$$

where $r_x(\tau)$ is the correlation coefficient envelope of the process x(t), ω_0 is the central frequency of the transmitted signal spectrum. Equation (23.12) may be rewritten in the more compact form

$$R_{\mathrm{r}}(\tau) = r_{\mathrm{r}}(\tau)\, r_x(\tau) \cos \omega_0 \tau,$$
$$r_{\mathrm{rN}}(\tau) = r_{\mathrm{r}}(\tau)\, r_x(\tau), \qquad \Big\} \qquad (23.13)$$

where $r_{\mathrm{rN}}(\tau)$ is the correlation coefficient envelope of the reverberation, $r_{\mathrm{r}}(\tau)$ is the correlation coefficient of the function $s_0(t)$, and is equal to:

$$r_{\mathrm{r}}(\tau) = (1/\delta_{\mathrm{ef}}) \int_{-\infty}^{\infty} s_0(t)\, s_0(t+\tau)\, dt. \qquad (23.14)$$

It is clear that Eqs. (23.13) and (23.14) may be used, after passage to the limit, to obtain the correlation characteristics of reverberation for the case when the transmitted signals have a sinusoidal carrier. This limiting transition corresponds to degeneration of the noise spectrum into a delta function. In this case,

$$r_x(\tau) \to 1, \qquad r_{\mathrm{rN}}(\tau) \to r_{\mathrm{r}}(\tau),$$

i.e., the reverberation correlation, as seen from (23.13), coincides with the relations obtained in Sec. 22 and is uniquely determined by the form of the function $s_0(t)$.

A noisy carrier is usually generated by the filtering of white (or very wide-band) noise by means of linear systems described by transfer constants $K(\omega)$. At the output of such a system (see [1, 12, 19]) the noise autocorrelation coefficient is defined by the expression

$$R_x(\tau) = \frac{\displaystyle\int_{-\infty}^{\infty} |K(\omega)|^2 \cos \omega\tau \, d\omega}{\displaystyle\int_{-\infty}^{\infty} |K(\omega)|^2 \, d\omega}, \qquad (23.15)$$

where the integral in the denominator is equal to the variance σ_x^2 of the process x(t), i.e.,

$$\sigma_x^2 = \langle x^2(t) \rangle = \int_{-\infty}^{\infty} |K(\omega)|^2 \, d\omega.$$

We will assume that the spectrum of the process x(t) has peaks at the frequencies $\omega = \omega_0$ and $\omega = -\omega_0$, and that

Table 23.1. Correlation Coefficient Envelopes for a Noisy Carrier

Filter characteristic $|K_1(\omega)|$	$r_x(\tau)$
Rectangular $1, \quad |\omega| \leqslant \Delta\omega_f/2$	$\dfrac{\sin(\Delta\omega_f \tau/2)}{\Delta\omega_f \tau/2}$
Bell-shaped $\exp[-(\omega/\Delta\omega_f)^2]$	$\exp(-\Delta\omega_f^2 \tau^2/4)$
Single section $[1 + (\omega/\Delta\omega_f)^2]^{-1/2}$	$\exp(-\Delta\omega_f |\tau|)$

$$|K(\omega_0)| = |K(\omega)|_{\max} = 1.$$

Then, with regard for Eq. (23.15) and the quasi-harmonicity of the given process, we obtain

$$R_x(\tau) = \frac{\int\limits_{-\omega_0}^{\infty} |K(\omega - \omega_0)|^2 \cos \omega\tau \, d\omega}{\int\limits_{-\omega_0}^{\infty} |K(\omega - \omega_0)|^2 \, d\omega} \cos \omega_0\tau.$$

Assuming also that the filter characteristic $K(\omega)$ is symmetrical with respect to the frequency $\omega = \omega_0$, we have for $r_X(\tau)$

$$r_x(\tau) = \frac{\int\limits_{0}^{\infty} |K_1(\omega)|^2 \cos \omega\tau \, d\omega}{\int\limits_{0}^{\infty} |K_1(\omega)|^2 \, d\omega}, \quad |K_1(\omega)| = |K(\omega - \omega_0)|. \tag{23.16}$$

Consequently, the reverberation correlation coefficient may be written in the following form on the basis of (23.13) and (23.16):

$$R_r(\tau) \approx \frac{\int\limits_{0}^{\infty} |K_1(\omega)|^2 \cos \omega\tau \, d\omega}{\int\limits_{0}^{\infty} |K_1(\omega)|^2 \, d\omega} r_r(\tau) \cos \omega_0\tau. \tag{23.17}$$

The results of calculating the correlation coefficient envelopes $r_X(\tau)$ of a noisy carrier $x(t)$ for various forms of the frequency characteristics $|K_1(\omega)|$ are shown in Table 23.1.

It is required, in order to calculate the reverberation correlation coefficient $R_r(\tau)$ for various forms of the characteristic $|K_1(\omega)|$ and forms of the function $s_0(t)$ to make use of the data in Tables 22.1 and 23.1, as well as Eq. (23.13) or the analogous relation (23.17).

An analysis of the resultant relations shows that for the transmission of noisy signals the correlation characteristics of the reverberation depend in general on the form of the function $s_0(t)$ and the form of the frequency characteristic $|K_1(\omega)|$ of the filter, as was the case in using signals with frequency modulation. This result is quite reasonable, insofar as, in general cases, reverberation represents a noisy stochastic process.

Hence, if, for example, we estimate the reverberation correlation interval for $s_0(t)$ and $|K_1(\omega)|$ determined in accordance with Eqs.(10.7) and row 3 of Table 23.1, then from the relation

$$\tau_r \approx \int\limits_{0}^{\delta} (1 - \tau/\delta) \exp(-\Delta\omega_f\tau) \, d\tau$$

we obtain

$$\tau_r \approx (1/\Delta\omega_f^2\delta) [\exp(-\Delta\omega_f\delta) + \Delta\omega_f\delta - 1]. \tag{23.18}$$

Given the condition

$$\Delta F_f \delta \ll 1, \quad \Delta\omega_f = 2\pi\Delta F_f, \tag{23.19}$$

i.e., for a very narrow filter band, we have for the correlation interval

$$\tau_{\mathrm{r}} \approx \delta / 2,$$ (23.20)

which agrees with the data of Table 22.1 (row 1).

If, on the other hand,

$$\Delta F_{\mathrm{f}} \delta \gg 1,$$ (23.21)

then

$$\tau_{\mathrm{r}} \approx 1 / \Delta \omega_{\mathrm{f}}.$$ (23.22)

Consequently, for the transmission of both frequency-modulated signals and signals with a noisy carrier, the reverberation correlation interval is described by the analogous relations (23.10) and (23.22).

§24. General Relations for the Correlation Characteristics of Envelope Fluctuations

The problem of calculating the correlation function of reverberation envelope fluctuations is solved with the following assumptions:

The distribution of instantaneous values of the reverberation signals is normal.

The reverberation constitutes a quasi-harmonic stochastic process.

Stationarization of the reverberation is performed over a sufficiently long interval $(-T/2, T/2)$, well in excess of the correlation interval $(T \gg \tau_r)$.

Under these assumptions, the analysis of the correlations in the reverberation envelope reduces to the analogous problem of investigating the correlation characteristics of the envelope of normal stationary stochastic processes.

The correlation coefficient of the reverberation envelope, being equal by definition to

$$R_{\mathrm{r}E}(\tau) = \frac{\langle E(t) E(t+\tau) \rangle - \langle E(t) \rangle^2}{\langle E^2(t) \rangle - \langle E(t) \rangle^2},$$ (24.1)

is calculated as follows. *

The four-dimensional probability density $W(E, E_\tau, \Psi, \Psi_\tau)$ of the reverberation envelope, and phase is found on the basis of the joint four-dimensional distribution $W_V(V_c, V_s, V_{c\tau}, V_{s\tau})$ of the quadrature components of their instantaneous values (see Sec. 16):

$$W(E, E_\tau, \Psi, \Psi_\tau) = W_V(E \cos \Psi, E \sin \Psi, E_\tau \cos \Psi_\tau, E_\tau \sin \Psi_\tau) |D_4|,$$
$$E, E_\tau \geqslant 0, \quad 0 \leqslant \Psi, \quad \Psi_\tau < 2\pi,$$ (24.2)

where $|D_4|$ is the modulus of the Jacobian transformation

$$D_4 = \frac{\partial (V_c, V_s, V_{c\tau}, V_{s\tau})}{\partial (E, E_\tau, \Psi, \Psi_\tau)},$$ (24.3)

---

* The method given in the present section for calculation of the correlation coefficient is similar to that described, for example, in [1, 12].

where

$$E = E(t), \quad E_\tau = E(t + \tau), \\ \Psi = \Psi(t), \quad \Psi_\tau = \Psi(t + \tau).$$ (24.4)

Considering the fact that in the given situation

$$D_4 = EE_\tau,$$

we obtain for the probability density $W(E, E_T, \Psi, \Psi_T)$ according to (24.2)

$$W(E, E_\tau, \Psi, \Psi_\tau) = EE_\tau W_V(E \cos\Psi, E \sin\Psi, E_\tau \cos\Psi_\tau, E_\tau \sin\Psi_\tau).$$ (24.5)

We then make use of the general expression (13.3) to find the four-dimensional normal law $W_V(V_C, V_S, V_{CT}, V_{ST})$, making the substitution therein according to

$$V_c = E \cos\Psi, \qquad V_s = E \sin\Psi, \\ V_{c\tau} = E_\tau \cos\Psi_\tau, \qquad V_{s\tau} = E_\tau \sin\Psi_\tau.$$

The correlation matrix K in this case takes the form

$$K = \begin{Vmatrix} 1 & 0 & r_\mathrm{I}(\tau) & 0 \\ 0 & 1 & 0 & r_\mathrm{I}(\tau) \\ r_\mathrm{I}(\tau) & 0 & 1 & 0 \\ 0 & r_\mathrm{I}(\tau) & 0 & 1 \end{Vmatrix},$$ (24.6)

where $r_\mathrm{r}(\tau)$ is the correlation coefficient envelope of the instantaneous reverberation values.

The determinant of the matrix (24.6) and its signed minors are expressed as follows:

$$|K| = [1 - r_\mathrm{I}^2(\tau)]^2, \\ |K_{11}| = |K_{22}| = |K_{33}| = |K_{44}| = 1 - r_\mathrm{I}^2(\tau), \\ |K_{12}| = |K_{21}| = |K_{14}| = |K_{41}| = |K_{23}| = |K_{32}| = |K_{34}| = 0, \\ |K_{13}| = |K_{31}| = |K_{24}| = |K_{42}| = -r_\mathrm{I}(\tau)[1 - r_\mathrm{I}^2(\tau)].$$ (24.7)

Taking the relations (13.3), (24.6), and (24.7) into account, the distribution (24.5) is written in the form

$$W(E, E_\tau, \Psi, \Psi_\tau) = \frac{EE_\tau}{4\pi^2\sigma_V^4[1 - r_\mathrm{I}^2(\tau)]} \exp\left\{ -\frac{E^2 + E_\tau^2 - 2EE_\tau r_\mathrm{I}(\tau)\cos(\Psi_\tau - \Psi)}{2\sigma_V^2[1 - r_\mathrm{I}^2(\tau)]} \right\}.$$ (24.8)

The four-dimensional probability density (24.8) makes it possible, by means of the relations

$$W(E, E_\tau) = \int_0^{2\pi}\int_0^{2\pi} W(E, E_\tau, \Psi, \Psi_\tau)\, d\Psi\, d\Psi_\tau, \quad E, E_\tau \geqslant 0,$$ (24.9)

$$W(\Psi, \Psi_\tau) = \int_0^\infty\int_0^\infty W(E, E_\tau, \Psi, \Psi_\tau)\, dE\, dE_\tau, \quad 0 \leqslant \Psi, \Psi_\tau < 2\pi,$$ (24.10)

to find the two-dimensional probability densities of the reverberation envelope and phase.

The distributions $W(E, E_\tau)$ and $W(\Psi, \Psi_\tau)$ may be used in particular to find the autocorrelation functions of the reverberation envelope $E(t)$ and phase $\Psi(t)$.

Substitution of Eq. (24.8) into (24.9) and integration yield the following relation for the two-dimensional probability density of the envelope:

$$W(E, E_\tau) = \frac{EE_\tau}{\sigma_V^4 [1 - r_\mathrm{I}^2(\tau)]} \exp\left\{ -\frac{E^2 + E_\tau^2}{2\sigma_V^2 [1 - r_\mathrm{I}^2(\tau)]} \right\} I_0 \left\{ \frac{r_\mathrm{I}(\tau) EE_\tau}{\sigma_V^2 [1 - r_\mathrm{I}^2(\tau)]} \right\},$$ (24.11)

where $I_0(X)$ is the zero-th order Bessel function of an imaginary argument (18.10).

The product mean $<EE_\tau>$ is defined as

$$\langle EE_\tau \rangle = \int\limits_0^\infty \int\limits_0^\infty EE_\tau W(E, E_\tau)\, dE\, dE_\tau.$$ (24.12)

Upon calculating $<EE_\tau>$ according to Eqs. (24.11) and (24.12), we arrive at complete elliptic integrals [1, 9], which may be written in series form. We thereby obtain for $<EE_\tau>$

$$\langle EE_\tau \rangle = (\pi/2)\, \sigma_V^2\, [1 + (1/4)\, r_\mathrm{I}^2(\tau) + (1/64)\, r_\mathrm{I}^4(\tau) + (1/2304)\, r_\mathrm{I}^6(\tau) + \ldots].$$ (24.13)

From this may be found, in particular, $<E>^2$, $<E^2>$, and the correlation coefficient $R_{\mathrm{r}E}(\tau)$ of the reverberation envelope.

Letting $\tau \to \infty$ and $\tau \to 0$, respectively, we obtain

$$\langle E \rangle^2 = (\pi/2)\, \sigma_V^2, \quad \langle E^2 \rangle = 2\sigma_V^2.$$ (24.14)

For $R_{\mathrm{r}E}(\tau)$ we have, on the basis of (24.1), (24.13), and (24.14),

$$R_{\mathrm{r}E}(\tau) \approx 0.91 r_\mathrm{I}^2(\tau) + 0.058 r_\mathrm{I}^4(\tau) + 0.0014 r_\mathrm{I}^6(\tau) + \ldots.$$ (24.15)

If we retain only the first term of the expansion (24.15), then

$$R_{\mathrm{r}E}(\tau) \approx r_\mathrm{I}^2(\tau),$$ (24.16)

which gives an error no greater than 10% in determining the correlation coefficient. Using the first two terms of the series leads to the relation

$$R_{\mathrm{r}E}(\tau) \approx 0.91 r_\mathrm{I}^2(\tau) + 0.09 r_\mathrm{I}^4(\tau);$$ (24.17)

now the error turns out to be less than 1%.

Equation (24.16) is sufficiently simple and in many cases involving the analysis of the correlation characteristics of the reverberation envelope its accuracy is quite satisfactory.

It turns out, therefore, that <u>the correlation coefficient of the reverberation envelope fluctuations is approximately equal to the square of the correlation coefficient envelope of its instantaneous values.</u>

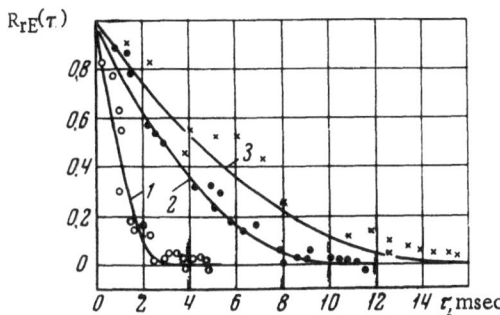

Fig. 30. Comparison of the measured correlation coefficients of the reverberation envelope fluctuations with the analytical (solid curves) for the transmission of rectangular pulses. 1) $\delta = 3$ msec; 2) $\delta = 10$ msec; 3) $\delta = 15$ msec.

§ 25. Correlation of Envelope Fluctuations for Various Forms of Transmitted Signals

The very simple expression (24.16), which relates the correlation coefficient for fluctuations of the reverberation envelope to the correlation coefficient envelope of its instantaneous values, enables one to find the values of $R_{rE}(\tau)$ at once on the basis of the results obtained in Secs. 22 and 23.

Below we examine the correlation characteristics of the reverberation envelope for the investigation of various types of signals and present some of the results of experimental investigations.

If the investigated pulses have a rectangular envelope and sinusoidal carrier, we obtain, on the basis of the data in Table 22.1 (row 1) and Eq. (24.16),

$$R_{rE}(\tau) \approx (1 - |\tau|/\delta)^2, \qquad |\tau| \leqslant \delta. \qquad (25.1)$$

We see that the correlation of the reverberation envelope fluctuations is uniquely determined by the duration of the transmitted pulses. We note that Eq. (25.1) is presented in [25], wherein the so-called time coherence intervals of the reverberation envelope are analyzed.

To illustrate the correspondence of the measured ocean data with the results of a calculation according to Eq. (25.1), Fig. 30 shows the correlation characteristics of the envelope of combined surface and bottom reverberation. The experimental data were processed according to the stationarized reverberation time intervals, including 7-10 independent readings, after which averaging was performed over 25 reverberation occurrences obtained from successively transmitted signals.

As apparent from the graph, the measured values of the correlation coefficients are near the analytical.

Of interest are the experimental data on the correlation characteristics of the reverberation envelope in the transmission of frequency-modulated pulses. Oscillograms showing bottom reverberation time intervals 15 msec in length for the transmission of rectangular pulses with different frequency deviations are presented in Fig. 31; the frequency variation followed a linear law.

It is apparent from these oscillograms that when the frequency deviation increases, the reverberation envelope fluctuates more rapidly. This indicates a reduction in the correlation interval and is consistent with the results of Sec. 23, in which it was remarked, in particular, that when the condition (23.4) is fulfilled, the correlation characteristics of the reverberation are determined by the law of frequency deviation. In fact, it can be shown that in this case the following relation holds for the correlation interval τ_{rE} of the reverberation envelope:

$$\tau_{rE} \approx k/\Delta F_M, \qquad 2\pi\Delta F_M = \Delta\omega_M, \qquad (25.2)$$

where k is a constant depending on the modulation law and form of the envelope, ΔF_M is the frequency deviation.

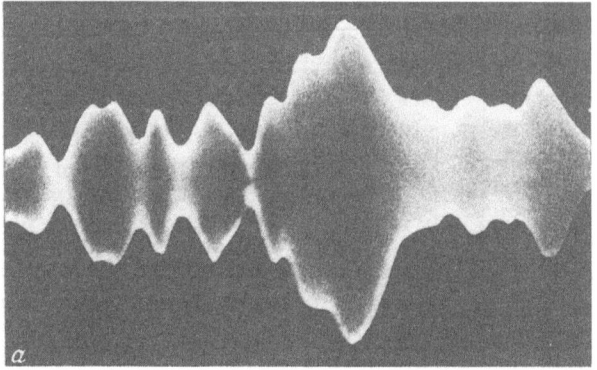

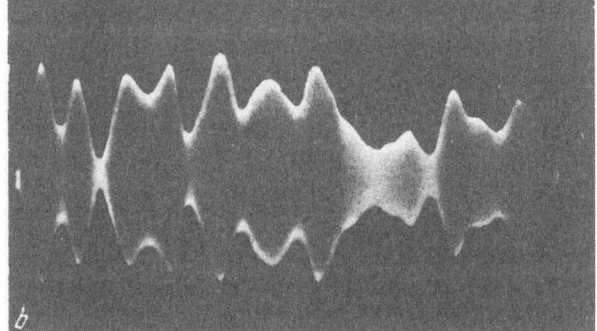

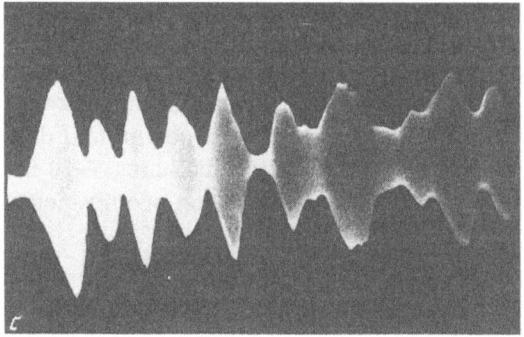

Fig. 31. Oscillograms of reverberation intervals for
various relative frequency deviations. a) $\Delta F_M \delta = 0$;
b) $\Delta F_M \delta = 2$; c) $\Delta F_M \delta = 3$; $f_0 = 60$ kc, $\delta = 3$ msec.

     An interesting observation is the fact that if the scatterers in the ocean medium are few,
or if there exists but a single scatterer (reflector), the character of the dependence (25.2) is
disrupted. This is illustrated by the data of Fig. 32, which shows the dependence of the correlation intervals on the frequency deviation for two cases: bottom reverberation and a signal
reflected from a compact obstacle having two or three effective scattering centers. The correlation interval is calculated from the reverberation oscillograms as the average fluctuation
period of its envelope relative to its mean value, where each processing series includes 10-15
independent time readings within the limits of the stationary reverberation interval. The averaged data (circles) were obtained in the processing of 20 reverberation events.

     It is evident from the data shown in Fig. 32 that for reverberation the correlation interval
is described by the relation (25.2), whereas, for a small number of scatterers, the envelope
begins to fluctuate more rapidly at larger values of the parameter $\Delta F_M \delta$.

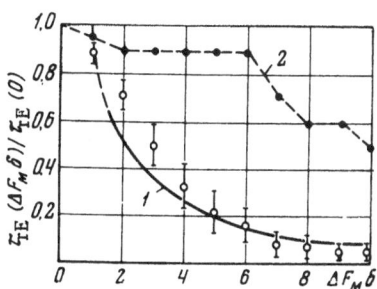

Fig. 32. Dependence of the measured normalized correlation interval for fluctuations of the reverberation envelope (empty circles) and the envelope of a signal reflected from a compact obstacle (dark circles, curve 2) on the relative frequency deviation; curve 1 corresponds to the law (25.2) for k = 1.

This behavior on the part of the reverberation envelope fluctuations confirms our postulated model of this process as a summation of a large number of stochastic perturbations.

We recall that the distribution of the reverberation envelope (see Sec. 18) is described by a Rayleigh law for any values of the parameter $\Delta F_M \delta$, so that its relative fluctuations do not vary as a function of the frequency deviation.

The independence of the correlation properties of the reverberation envelope from the pulse duration is also observed in the investigation of noisy signals. Surface reverberation oscillograms obtained in the transmission of pulses with a sinusoidal and a noisy carrier are shown for comparison in Fig. 33, where, in the second case, the fluctuations of the envelope are determined only by the frequency band of the carrier.

Data on the correlation characteristics of the reverberation envelope for noisy signals are shown in Fig. 34; here, since the filter shaping the noisy carrier had a frequency characteristic in the form of a single-section resonance curve, while the signals represented a noise segment of duration δ, the analytical correlation coefficient was assumed on the basis of row 3 of Table 23.1 and Eq. (24.16) to be equal to

$$R_{rE}(\tau) \approx \exp\left(-2\Delta\omega_f |\tau|\right)(1 - |\tau|/\delta)^2, \quad |\tau| \leqslant \delta. \tag{25.3}$$

Under the condition (23.21) the following relation applies to $R_{rE}(\tau)$:

$$R_{rE}(\tau) \approx \exp\left(-2\Delta\omega_f |\tau|\right), \tag{25.4}$$

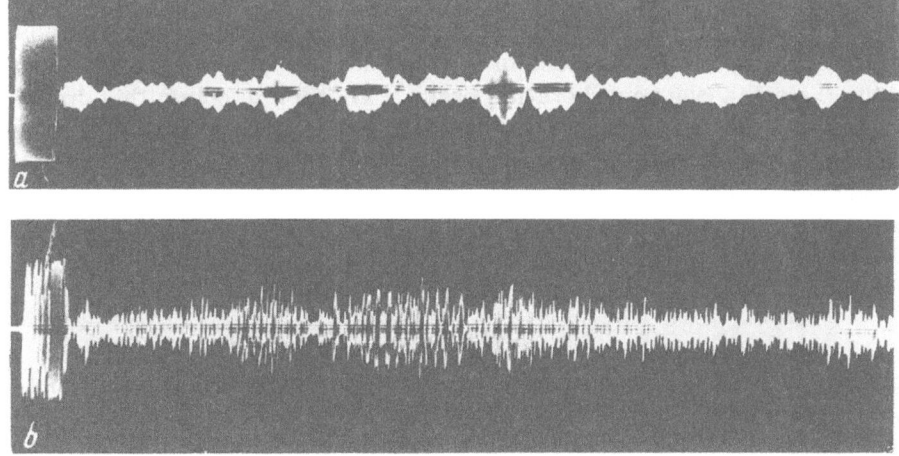

Fig. 33. Oscillograms of stationarized reverberation. a) For a sinusoidal carrier; b) for a noisy carrier with $\Delta F_N = 2.1$ kc, $f_0 = 15$ kc, $\delta = 10$ msec.

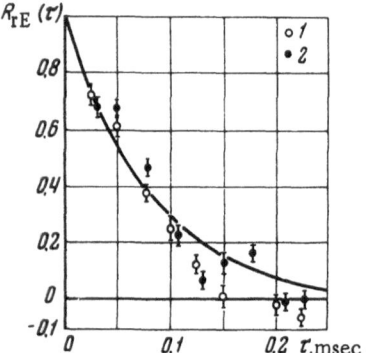

Fig. 34. Measured and calculated (curve) autocorrelation coefficient for fluctuations of the reverberation envelope in the transmission of noisy signals. 1) For δ = 5 msec; 2) δ = 20 msec; f_0 = 36 kc, ΔF_N = 1 kc.

from which it follows that the correlation of the envelope does not depend on the duration of the transmitted signal.

We next examine data on the correlation characteristics of the reverberation envelope for the transmission of a pulse train. It is evident from the reverberation oscillograms (Fig. 35) that the fluctuations of the envelope in this case have a dual character; fluctuations with a mean period of the same order as the total signal duration are noted along with pronounced fluctuations with a period of the same order as the duration of an individual pulse within the train.

Data on the envelope correlation coefficient for the transmission of such signals are shown in Fig. 36. It follows from the figure that the analytical values of $R_{rE}(\tau)$ provide a satisfactory fit for the measured data.

We presented above the results of a correlation analysis of reverberation signals produced by the transmission of relatively short pulses. For long pulses the theoretical principles may be violated due to motion of the scatterers and acoustic arrays, and in some cases due to departure of the scatterer distribution in the ocean from a Poisson distribution. In particular, a lack of correspondence between the experimental and analytical data is observed at times for surface reverberation when scattering is caused by air bubbles in motion as the result of the wave state of the ocean surface (see also Sec. 44).

The measured values of the correlation coefficient of reverberation envelope fluctuations are shown in Fig. 37, where they exhibit considerable deviation from the relation (25.1). The observed periodicity of the correlation curve is attributed, clearly, to surface waves on the ocean, the state of which was characterized by 2-3 points on the sea height scale at the time of the given measurements.

§ 26. Instantaneous Phase Correlation

The general expression (24.10) makes it possible to calculate the two-dimensional probability density $W(\Psi, \Psi_\tau)$, by which it is possible, in particular, to find the autocorrelation characteristics of the instantaneous reverberation phase. In order to obtain this distribution, it is necessary to substitute the relation (24.8) for the four-dimensional probability density of the envelope and phase into (24.10) and to carry out double integration. Such a calculation (see [1]) leads to the following result:

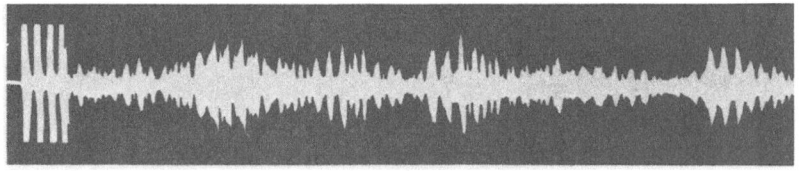

Fig. 35. Oscillograms of stationarized reverberation in the transmission
of a train of four pulses.

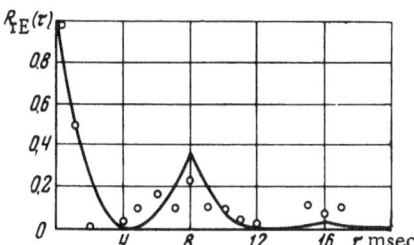

Fig. 36. Measured (circles) and calculated (curve) autocorrelation coefficients for fluctuations of the reverberation envelope in the transmission of a train of three rectangular pulses. $\delta_1 = 4$ msec, $T_0 = 8$ msec.

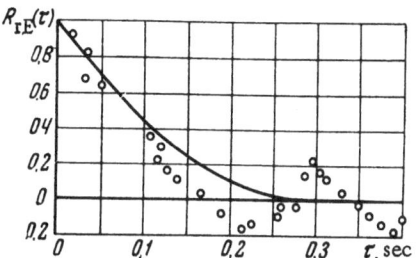

Fig. 37. Measured autocorrelation coefficient for fluctuations of the surface reverberation envelope. $f_0 = 15$ kc, $\delta = 0.3$ sec.

$$W(\Psi, \Psi_\tau) = (1/4\pi^2)[1 - r_{\mathrm{I}}^2(\tau)]\left\{\frac{1}{1 - r_{\mathrm{I}}^2(\tau)\cos^2(\Psi - \Psi_\tau)} + \right.$$

$$\left. + r_{\mathrm{I}}(\tau)\cos(\Psi - \Psi_\tau)\frac{(\pi/2) + \arcsin[r_{\mathrm{I}}(\tau)\cos(\Psi - \Psi_\tau)]}{[1 - r_{\mathrm{I}}^2(\tau)\cos^2(\Psi - \Psi_\tau)]^{1/2}}\right\},$$

$$0 \leqslant \Psi,\ \Psi_\tau < 2\pi, \tag{26.1}$$

where, as before, $r_{\mathrm{I}}(\tau)$ is the correlation coefficient envelope of the instantaneous reverberation values. We note that the given process is assumed to be quasi-harmonic.

Equation (26.1) enables one to calculate the moments of the two-dimensional phase distribution, including the correlation coefficient

$$R_{\mathrm{I}\Phi}(\tau) = \frac{\langle\Psi(t)\Psi(t+\tau)\rangle - \langle\Psi(t)\rangle^2}{\langle\Psi^2(t)\rangle - \langle\Psi(t)\rangle^2} \tag{26.2}$$

of the fluctuations in the instantaneous reverberation phase.

It is interesting that $W(\Psi, \Psi_T)$ depends only on the phase difference, and the one-dimensional distributions $W(\Psi)$ and $W(\Psi_T)$ are uniform:

$$W(\Psi) = 1/2\pi,\ W(\Psi_\tau) = 1/2\pi. \tag{26.3}$$

The product mean

$$\langle\Psi\Psi_\tau\rangle = \int\limits_0^{2\pi}\int\limits_0^{2\pi}\Psi\Psi_\tau W(\Psi, \Psi_\tau)\,d\Psi\,d\Psi_\tau \tag{26.4}$$

is expressed in the following series form after substitution of Eq. (26.1) into (26.4) and integration [12]:

$$\langle\Psi\Psi_\tau\rangle = \pi^2\left[1 + (1/\pi)\arcsin r_{\mathrm{I}}(\tau) - (1/\pi^2)\arcsin^2 r_{\mathrm{I}}(\tau) + (0.5/\pi^2)\sum_{k=1}^{\infty}\frac{r_{\mathrm{I}}^{2k}(\tau)}{k^2}\right]. \tag{26.5}$$

From this we find

$$\langle\Psi\rangle = \pi,\quad \langle\Psi^2\rangle = (4/3)\pi^2. \tag{26.6}$$

The correlation coefficient $R_{r\varphi}(\tau)$ is determined in accordance with (26.2), (26.5), and (26.6) by the relation

$$R_{r\Phi}(\tau) = (3/\pi) \, \text{arc sin} \, r_r(\tau) - (3/\pi^2) \, \text{arc sin}^2 r_r(\tau) +$$
$$+ (3/2\pi^2) \, r_r^2(\tau) + (3/8\pi^2) \, r_r^4(\tau) + (1/6\pi^2) \, r_r^6(\tau) + \ldots \qquad (26.7)$$

It is permissible for approximate calculations to use the approximate expression

$$R_{r\Phi}(\tau) \approx (3,2/\pi) \, \text{arc sin} \, r_r(\tau) - (3,2/\pi^2) \, \text{arc sin}^2 r_r(\tau) + 0,2 r_p(\tau) \qquad (26.8)$$

or the simpler relation

$$R_{r\Phi}(\tau) \approx (2/\pi) \, \text{arc sin} \, r_r(\tau), \qquad (26.9)$$

although the latter yields considerable error over the precise formula (26.7) for certain values of $r_r(\tau)$.

The above relations may be used to calculate the correlation coefficients $R_{r\varphi}(\tau)$ of the time fluctuations of the instantaneous reverberation phase if the function $r_r(\tau)$ is known. For this it is necessary to use the data in Secs. 22 and 23 on the reverberation correlation coefficient envelope, corresponding to various types of transmitted signals.

We note that the autocorrelation characteristics for the reverberation envelope and phase are generally quite similar to one another in their behavior.

§ 27. Spatial Correlation

In this section we investigate the following problem: to find the cross-correlation function of reverberation signals received by two point receivers placed at some distance apart for a specified directivity characteristic for the transmitting array.

We analyze the spatial correlation of the reverberation for various geometric arrangements of the scattering regions in the ocean.

We first consider the case of sound scattering by inhomogeneities concentrated in a fairly thin layer, when the transmitters and receivers are situated in a layer itself or in its immediate vicinity. The process generated as a result of scattering by inhomogeneities of the layer itself provide a good description of surface or bottom reverberation. Volume reverberation caused, for example, by scattering at inhomogeneities of sufficiently localized scattering layers are also described by such a model.

For the analysis of the spatial cross-correlation of the reverberation signals we make use of a generalization of the theorem of superposition of stochastic perturbations (Sec. 7), introducing the appropriate dependences of the quantities a_i and t_i on the spatial coordinates.

We investigate two processes:

$$\left.\begin{array}{l} V_1(t) = \sum_{i=0}^{\infty} a_i(\alpha_i) \, s\,[t - t_i - \Delta t(\alpha_i)/2], \\[2mm] V_2(t) = \sum_{j=0}^{\infty} a_j(\alpha_j) \, s\,[t - t_j - \Delta t(\alpha_j)/2], \end{array}\right\} \qquad (27.1)$$

which determine the reverberation signals at two receivers separated by a distance r (Fig. 38). As before, the function s(t) describes the form of the transmitted signals. It is inferred from the generalization of the theorem of superposition of stochastic perturbations that, in order to calculate the spatial cross-correlation function of the reverberation, it is necessary to specify

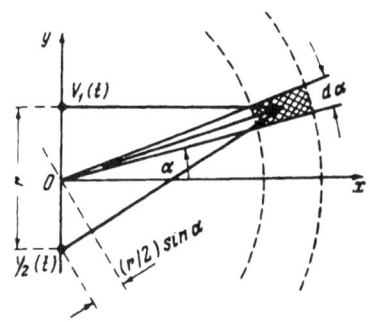

Fig. 38. Diagram used in calculating the spatial correlation of reverberation produced by scattering at inhomogeneities concentrated in a thin layer.

the probability density of the stochastic parameters on which the form of the perturbations $s_i(t)$ and $s_j(t)$ depends. Such a parameter in the given instance is the angle α_i in terms of which is expressed the quantity $\Delta t(\alpha_i)$ characterizing the random path difference of the i-th elementary scattered signal arriving at the two receivers.

If we assume the indicated probability density to be equal to $W(\alpha)$, the spatial cross-correlation function

$$B_{\text{rS}}(r) = \langle V_1(t)V_2(t)\rangle$$

may be written in the following form, on the basis of Eqs. (7.3) and (27.1):

$$B_{\text{rS}} = \langle n_1\rangle \int_0^{2\pi} \langle a^2(\alpha)\rangle W(\alpha) \int_{-\infty}^{\infty} s[t - \Delta t(\alpha)/2]\, s[t + \Delta t(\alpha)/2]\, dt\, d\alpha, \tag{27.2}$$

where $\langle a^2(\alpha)\rangle$ is a function describing the dependence of the intensity of the scattered signals on the angle α and is determined by the normalized directivity characteristic of the transmitter $\varphi_T(\alpha)$:

$$\langle a^2(\alpha)\rangle = \langle a^2\rangle\, \varphi_T^2(\alpha). \tag{27.3}$$

The stochastic variable $\Delta t(\alpha)$ is equal to

$$\Delta t(\alpha) = (r/c)\sin\alpha, \tag{27.4}$$

where c is the velocity of sound propagation.

We also take into account the fact that $\Delta t(\alpha) \ll \delta_{\text{ef}}$, as well as the condition of quasi-harmonicity of the transmitted signals, from which it follows that

$$\int_{-\infty}^{\infty} s[t - \Delta t(\alpha)/2]\, s[t + \Delta t(\alpha)/2]\, dt \approx (\delta_{\text{ef}}/2)\cos[\omega_0\Delta t(\alpha)],$$

where δ_{ef} is defined in (11.1). Then, bearing Eqs. (27.3) and (27.4) in mind, we go from (27.2) to the expression

$$B_{\text{rS}}(r) \approx (\langle n_1\rangle\langle a^2\rangle\delta_{\text{ef}}/2) \int_0^{2\pi} \varphi_T(\alpha) W(\alpha)\cos(kr\sin\alpha)\, d\alpha, \tag{27.5}$$

where

$$k = 2\pi/\lambda = \omega_0/c$$

is the wave number.

We also assume that the distribution of scatterers in the layer is uniform, i.e.,

$$W(\alpha) = 1/2\pi, \quad 0 \leqslant \alpha < 2\pi. \tag{27.6}$$

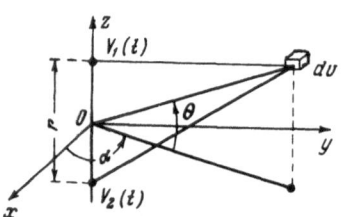

Fig. 39. Coordinate system used in calculating the spatial correlation of reverberation for volume scattering.

This type of distribution corresponds to a constant mean scatterer density in the layer.

Substituting Eq. (27.6) into (27.5), we obtain the following final expression for $B_{rS}(r)$:

$$B_{rS}(r) \approx (\langle n_1 \rangle \langle a^2 \rangle \, \delta_{ef}/4\pi) \int\limits_0^{2\pi} \varphi_T^2(\alpha) \cos(kr \sin \alpha) \, d\alpha. \qquad (27.7)$$

For the correlation coefficient $R_{rS}(\tau)$ we have, on the basis of (27.7),

$$R_{rS}(r) \approx (1/\Delta\alpha_{ef}) \int\limits_0^{2\pi} \varphi_T^2(\alpha) \cos(kr \sin \alpha) \, d\alpha, \qquad (27.8)$$

where the parameter

$$\Delta\alpha_{ef} = \int\limits_0^{2\pi} \varphi_T^2(\alpha) \, d\alpha \qquad (27.9)$$

determines the effective width of the directivity characteristic of the transmitting array.

Equation (27.8) provides a means for calculating the spatial correlation characteristics of reverberation for transmitting acoustic arrays with various directivity patterns.

We next consider the case when the scatterers are distributed throughout all space. Using the coordinate system shown in Fig. 39, and carrying out statistical averaging over all possible values of α and θ according to the general expression (7.3) for the correlation function $B_{rS}(r)$, we obtain

$$B_{rS}(r) = \langle n_1 \rangle \int\limits_{-\pi/2}^{\pi/2} \int\limits_0^{2\pi} \langle a^2(\alpha, \theta) \rangle W(\alpha, \theta) \int\limits_{-\infty}^{\infty} s[t - \Delta t(\alpha, \theta)/2] \times s[t + \Delta t(\alpha, \theta)/2] \, dt \, d\alpha \cos \theta \, d\theta, \quad (27.10)$$

where, in the given case,

$$\Delta t(\alpha, \theta) = (r/c) \sin \theta. \qquad (27.11)$$

Assuming, as before, that the scatterers are distributed independently throughout space and that their mean density in space is constant, we write

$$\left. \begin{array}{l} W(\alpha, \theta) = W(\alpha) W(\theta), \\ W(\alpha) = 1/2\pi, \quad 0 \leqslant \alpha < 2\pi, \\ W(\theta) = 1/\pi, \quad |\theta| \leqslant \pi/2. \end{array} \right\} \qquad (27.12)$$

We now bring in the normalized directivity characteristic of the transmitter, whereupon

$$\langle a^2(\alpha, \theta) \rangle = \langle a^2 \rangle \, \varphi_n^2(\alpha, \theta). \qquad (27.13)$$

On the basis of (27.10)-(27.13), recognizing the quasi-harmonicity of the transmitted signals, we obtain for the spatial correlation coefficient of the reverberation

$$R_{rS}(r) \approx (1/\Delta\Omega_{ef}) \int_{-\pi/2}^{\pi/2} \int_{0}^{2\pi} \varphi_T^2(\alpha, \theta) \cos(kr \sin\theta) \, d\alpha \cos\theta \, d\theta, \tag{27.14}$$

where $\Delta\Omega_{ef}$ denotes a parameter governing the spatial angle of divergence of the characteristic $\varphi_T(\alpha, \theta)$:

$$\Delta\Omega_{ef} = \int_{-\pi/2}^{\pi/2} \int_{0}^{2\pi} \varphi_T^2(\alpha, \theta) \, d\alpha \cos\theta \, a\theta. \tag{27.15}$$

For our subsequent calculations of the spatial cross-correlation characteristics of reverberation signals according to the relations (27.8) and (27.14), we need to know the specific form of the functions $\varphi_T(\alpha)$ and $\varphi_T(\alpha, \theta)$.

We indicate one special result that ensues from the relations obtained above and pertains to the transmission of signals by a nondirectional array. In this case, the directivity characteristic of the transmitter is described by the following expression for sound scattering by inhomogeneities situated in a thin layer:

$$\left.\begin{array}{l} \varphi_T(\alpha) = 1, \quad 0 \leqslant \alpha < 2\pi, \\ \Delta\alpha_{ef} = 2\pi. \end{array}\right\} \tag{27.16}$$

After substitution of (27.16) into (27.8) we obtain for the spatial correlation coefficient of the reverberation

$$R_{rS}(r) \approx J_0(kr), \tag{27.17}$$

where $J_0(X)$ is the zero-th order Bessel function of a real argument [9]:

$$J_0(X) = (1/\pi) \int_{0}^{\pi} \cos(X \sin\varphi) d\varphi. \tag{27.18}$$

It is apparent from Eq. (27.17) that the spatial correlation of reverberation due to scattering by inhomogeneities concentrated in a thin layer with nondirectional transmission is determined by the central frequency of the transmitted-signal spectrum or, accordingly, by the acoustic wavelength at this frequency.

If the scattering occurs at inhomogeneities distributed throughout the entire volume of the medium, and the signal transmission is produced as in the foregoing case, by a nondirectional array, so that

$$\begin{array}{l} \varphi_T(\alpha, \theta) = 1, \quad 0 \leqslant \alpha < 2\pi, \quad |\theta| \leqslant \pi/2, \\ \Delta\Omega_{ef} = 2\pi^2, \end{array} \tag{27.19}$$

then for the correlation coefficient we obtain, according to Eqs. (27.14) and (27.19),

$$R_{rS}(r) \approx \frac{\sin(kr)}{kr}. \tag{27.20}$$

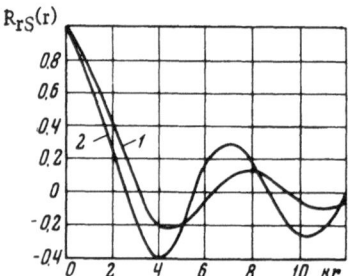

Fig. 40. Spatial correlation coefficients of reverberation for nondirectional transmission and scattering by inhomogeneities concentrated in an unbounded medium (1) and in a thin layer (2).

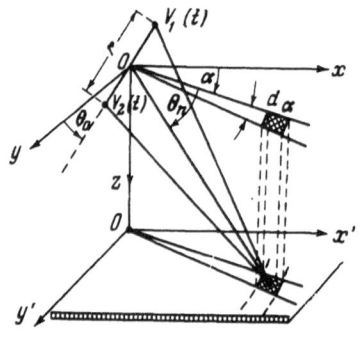

Fig. 41. Coordinate system used in calculating the spatial correlation of reverberation for irradiation of a thin layer at some angle.

Curves are presented in Fig. 40 for the correlation coefficients calculated according to Eqs. (27.17) and (27.20).

The described method of calculating the spatial correlation coefficients of reverberation permits one to solve other similar problems as well as involving various geometric arrangements of the scatterers in the ocean medium and various directivity characteristics for the acoustic arrays. In particular, the scatterer allocation shown in Fig. 41 is an important one and the most general, wherein a pair of receivers situated in the yz plane are rotated about the y axis through an angle θ_0, and the reverberation signals $V_1(t)$ and $V_2(t)$ are elicited by scattering at inhomogeneities concentrated in a thin unbounded layer situated in the plane x'y'. The analysis in this case refers, for example, to irradiation of the ocean surface or bottom at a certain angle.

Averaging over the ensemble of states of the product $<V_1(t)V_2(t)>$ for some fixed instant in time, and regarding the assumed coordinate system, we write the following general expression for the reverberation cross-correlation function:

$$B_{rS}(r) = \langle n_1 \rangle \langle a^2 \rangle \int_{-\pi/2}^{\pi/2} \int_0^{2\pi} \varphi_T^2(\alpha - \alpha_T, \theta - \theta_T) W(\alpha, \theta) \times$$

$$\times \int_{-\infty}^{\infty} s[t - \Delta t(\alpha, \theta)/2] \, s[t + \Delta t(\alpha, \theta)/2] \, dt \, d\alpha \cos\theta \, d\theta, \quad (27.21)$$

where $\varphi_T(\alpha - \alpha_T, \theta - \theta_T)$ is the normalized directivity characteristic of the transmitter, the axis of which forms an angle α_T with the xz plane and an angle θ_T with the xy plane. Recognizing that the layer is thin, it may be approximately considered that $\theta \approx \theta_T$. The value of the stochastic variable $\Delta t(\alpha, \theta)$ for times $t \gg \Delta t(\alpha, \theta)$ is determined in the form

$$\Delta t(\alpha, \theta) \approx (r/c)(\cos\theta_0 \cos\theta_T \sin\alpha + \sin\theta_0 \sin\theta_T). \quad (27.22)$$

Then, on the basis of (27.6), (27.21), and (27.22), taking into account the quasi-harmonicity of the transmitted signals, we obtain for the spatial correlation coefficient

$$R_{rS}(r) \approx (1/\Delta\alpha_{ef})[\cos(kr \sin\theta_0 \sin\theta_T) \int_0^{2\pi} \varphi_T^2(\alpha - \alpha_T) \times$$

$$\times \cos(kr \cos\theta_0 \cos\theta_T \sin\alpha) \, d\alpha - \sin(kr \sin\theta_0 \sin\theta_T) \times$$

$$\times \int_0^{2\pi} \varphi_T^2(\alpha - \alpha_T) \sin(kr \cos\theta_0 \cos\theta_T \sin\alpha) \, d\alpha], \quad (27.23)$$

where $\Delta\alpha_{ef}$ is defined by Eq. (27.9).

Equation (27.8) may be derived from (27.23) in particular, if we let $\theta_0 = 0$ and $\theta_T = 0$, i.e., reduce the problem to an investigation of the spatial correlation of reverberation in scattering by inhomogeneities of a thin layer situated in the plane xy (see Fig. 38).

For nondirectional transmission, allowing for (27.16) and the fact that

$$\int_0^{2\pi} \sin(A \sin \varphi)\, d\varphi = 0,$$

we obtain for $R_{rS}(r)$

$$R_{rS}(r) \approx J_0(kr \cos\theta_0 \cos\theta_T). \tag{27.24}$$

In the more general case of directional radiation we find the following relations from Eq. (27.23):

for $\theta_T = 0$:

$$R_{rS}(r) \approx (1/\Delta\alpha_{ef}) \int_0^{2\pi} \varphi_T^2(\alpha - \alpha_T) \cos(kr \cos\theta_0 \sin\alpha)\, d\alpha; \tag{27.25}$$

for $\theta_0 = 0$:

$$R_{rS}(r) \approx (1/\Delta\alpha_{ef}) \int_0^{2\pi} \varphi_T^2(\alpha - \alpha_T) \cos(kr \cos\theta_T \sin\alpha)\, d\alpha. \tag{27.26}$$

For subsequent calculations it is necessary to assign the specific form of the transmitter directivity characteristic and its orientation, and to use one of the above relations, depending on the geometric arrangement of the scatterers.

It is instructive to find an expression for the correlation coefficient for the case of a sharp transmitter directivity. If the scatterers are situated in a thin layer, and $\varphi_T(\alpha - \alpha_T)$ is an even function, then, taking into account the approximate equality $\sin\alpha \approx \alpha$ for small angles α, we find, for example, from (27.26):

$$R_{rS}(r) \approx (2/\Delta\alpha_{ef}) \int_0^{\infty} \varphi_T^2(\alpha - \alpha_T) \cos(kr \cos\theta_T \alpha)\, d\alpha. \tag{27.27}$$

Hence it is clear that the reverberation correlation coefficient over the wave front is the Fourier cosine transform of the square of the transmitter directivity characteristic.

Equation (27.27) is particularly useful in comparing the analytical relations with the experimental data on the spatial correlation of surface and bottom reverberation. We specify the directivity characteristic of the transmitter by the equation

$$\varphi_T(\alpha) = [1 + (\alpha/\Delta\alpha_T)^2]^{-1/2}, \quad |\alpha| \leqslant \pi, \quad \Delta\alpha_{ef} = \pi\Delta\alpha_T \ll 1, \tag{27.28}$$

which satisfactorily describes the directivity characteristic of typical acoustic arrays used in ocean measurements. Then from (27.27) and (27.28) we get, for $\alpha_T = 0$,

$$R_{rS}(r) \approx \exp(-\Delta\alpha_{ef} k \cos\theta_T |r|/\pi). \tag{27.29}$$

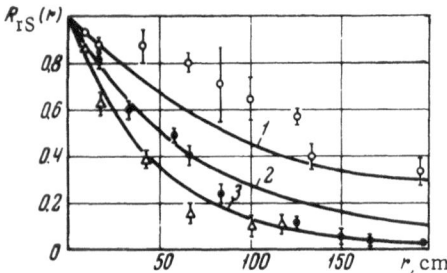

Fig. 42. Spatial correlation coefficients of surface reverberation for $\Delta\alpha_{ef} = 0.12$, corresponding to various signal carrier frequencies.   1) $f_0 = 4$ kc; 2) 7 kc; 3) 11 kc.

If we make further allowance for the fact that the quantity $\Delta\alpha_{ef}$ is determined by the transmitter dimensions d and wavelength λ, so that

$$\Delta\alpha_{ef} \approx \lambda/d, \qquad (27.30)$$

then, for example, with $\theta_T = 0$, we obtain for $R_{rS}(r)$

$$R_{rS}(r) \approx \exp(-2|r|/d). \qquad (27.31)$$

Consequently, the reverberation correlation coefficient over the wave front in this case is uniquely determined by the ratio of the distance between receivers to the size of the transmitter. If for different frequencies

$$\Delta\alpha_{ef} = \text{const},$$

then Eq. (27.31) is conveniently rewritten in the form

$$R_{rS}(r) \approx \exp(-2\Delta\alpha_{ef}|r|/\lambda). \qquad (27.32)$$

In order to check the correspondence of the theoretical results with the experimental data, measurements of the spatial correlation coefficient of surface reverberation were performed. The final results were obtained as the result of averaging over ten reverberation events, with correlation analysis run for each during the measurements. The instantaneous values of the correlation coefficients for each reverberation event were found by taking the time average of 20 to 30 independent readings by means of an electronic correlator operating on the principle of phase coincidences of the processes being correlated [4].

In one series of tests the directivity pattern of the transmitters in the horizontal plane was the same for all carrier frequencies and was determined by the parameter

$$\Delta\alpha_{ef} \approx 0.12.$$

The correlation coefficients calculated according to Eq. (27.32) and the corresponding experimental data are illustrated in Fig. 42.

In another series of measurements a single transmitter was used at all frequencies. The directivity characteristic of this transmitter was adequately approximated by an exponential squared, so that

$$\varphi_T(\alpha) \approx \exp[-(\alpha/\Delta\alpha_T)^2], \qquad \Delta\alpha_T \ll 1. \qquad (27.33)$$

Calculation of the spatial correlation coefficient from Eqs. (27.27) and (27.33) for $\alpha_T = 0$ and $\theta_T = 0$ yields the following result in this case:

$$R_{rS}(r) \approx \exp[-(kr\Delta\alpha_T)^2/4]. \qquad (27.34)$$

If we then take (27.30) into account, along with the fact that $\Delta\alpha_{ef} = \sqrt{\pi/2}\Delta\alpha_T$, we obtain from (27.34):

$$R_{rS}(r) \approx \exp[-\sqrt{2}(r/d)^2]. \qquad (27.35)$$

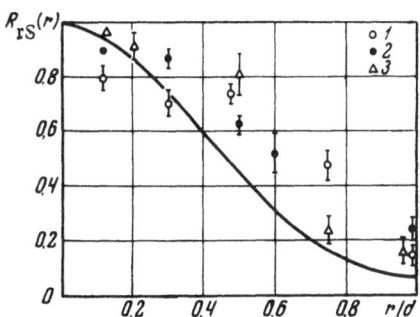

Fig. 43. Spatial correlation coefficients of surface reverberation for d = 80 cm. 1) f_0 = 3 kc; 2) 5 kc; 3) 7 kc.

The measured values of $R_{rS}(r)$ and a curve drawn in accordance with Eq. (27.35) are shown in Fig. 43.

The figures exhibit satisfactory agreement between the calculated dependences and the results of the ocean measurements.

An interesting observation is that results similar to some of those presented above are found in several papers (see [28, 32, 40]), in which certain aspects of the cross-correlation of noise fields generated by spatially distributed sources are investigated.

In particular, for example, in [40], the space—time correlation of summed signals received at two points and produced by a set of point sources distributed at random over a spherical and an annular surface is investigated. The problem is solved for a spectrum of arbitrary form on the part of the vibrations transmitted by the sources. Another interesting paper is [28], in which the results are presented from an investigation of the spatial correlation of sound fields in a reverberation chamber, where the measured correlation coefficients for the case of a diffuse field turned out to be close to a dependence of the type (27.20). Results appropriate to sea reverberation appear to have been first published in [13, 18].

§28. Correlation of Reverberation Signals Arriving from Different Directions

In the preceding section we investigated the spatial correlation characteristics of reverberation signals with sound reception by nondirectional hydrophones.

It is important also to study the correlation in the case when reception is realized by means of directional acoustic arrays.

We first examine the case when scattering occurs at inhomogeneities concentrated in a thin layer, in the immediate vicinity of which are located the acoustic arrays. Let the directivity characteristics of the receiving arrays in the horizontal plane be identical and described by the relations

$$\varphi_{R_1}(\alpha) = \varphi_R(\alpha + \Delta\alpha_0/2), \qquad \varphi_{R_2}(\alpha) = \varphi_R(\alpha - \Delta\alpha_0/2), \tag{28.1}$$

where $\Delta\alpha_0$ is the angle between the axial direction of the given characteristics, which are assumed to be normalized, so that

$$\varphi_{R_1}(0) = 1, \qquad \varphi_{R_2}(0) = 1. \tag{28.2}$$

We denote the normalized directivity characteristic of the transmitting array by $\varphi_T(\alpha)$.

Let us now compute the cross-correlation function of reverberation signals received by two arrays with the directivity characteristics defined in Eq. (28.1).

For the solution of this problem, we use, as in Sec. 27, the general relation (7.3). In the given instance, carrying out averaging over the random values of the angle α_i, we obtain for the cross-correlation function $B_{12}(\Delta\alpha_0)$

$$B_{12}(\Delta\alpha_0) = \langle n_1 \rangle \int_0^{2\pi} \langle a_1(\alpha)\,a_2(\alpha) \rangle W(\alpha) \int_{-\infty}^{\infty} s^2(t)\,dt, \qquad (28.3)$$

where $a_1(\alpha)$ and $a_2(\alpha)$ are the amplitudes of the elementary scattered signals, the product mean of which may be expressed in terms of the characteristics $\varphi_{R1}(\alpha)$, $\varphi_{R2}(\alpha)$, and $\varphi_T(\alpha)$ in the form

$$\langle a_1(\alpha)\,a_2(\alpha) \rangle = \langle a^2 \rangle\,\varphi_T^2(\alpha)\,\varphi_R(\alpha + \Delta\alpha_0/2)\,\varphi_R(\alpha - \Delta\alpha_0/2). \qquad (28.4)$$

We note that α is a stochastic parameter, the probability density of which is equal to $W(\alpha)$.

Assuming that scattering is caused by inhomogeneities that are statistically uniformly distributed in space, and taking into account the quasi-harmonicity of the function s(t) and Eqs. (11.1), (27.6), and (28.3)-(28.5), we obtain for $B_{12}(\Delta\alpha_0)$

$$B_{12}(\alpha_0) \approx (\langle n_1 \rangle \langle a^2 \rangle\, 0_{ef}/4\pi) \int_0^{2\pi} \varphi_T^2(\alpha)\,\varphi_R(\alpha + \Delta\alpha_0/2)\,\varphi_R(\alpha - \Delta\alpha_0/2)\,d\alpha. \qquad (28.5)$$

For the cross-correlation coefficient $R_{12}(\Delta\alpha_0)$ we obtain from (28.5)

$$R_{12}(\Delta\alpha_0) \approx (1/\Delta\alpha_{TRef}) \int_0^{2\pi} \varphi_T^2(\alpha)\,\varphi_R(\alpha + \Delta\alpha_0/2)\,\varphi_R(\alpha - \Delta\alpha_0/2)\,d\alpha, \qquad (28.6)$$

where

$$\Delta\alpha_{TRef} = \int_0^{2\pi} \varphi_T^2(\alpha)\varphi_R^2(\alpha)\,d\alpha \qquad (28.7)$$

is a parameter accounting for the shape of the directivity pattern of the transmitting and receiving arrays.

It follows from Eq. (28.6) that the cross-correlation of reverberation signals arriving from different directions is uniquely determined by the form of the directivity characteristics $\varphi_T(\alpha)$ and $\varphi_R(\alpha)$.

As an example, we analyze the case when the transmission is nondirectional, i.e., $\varphi_T(\alpha) = 1$, while the directivity of the receiving arrays is determined by a function of the form

$$\varphi_R(\alpha) = \exp\left[-(\pi/2)\,(\alpha/\Delta\alpha_{Ref})^2\right], \qquad |\alpha| \leqslant \pi, \qquad (28.8)$$

where, in the given case,

$$\Delta\alpha_{Ref} = \Delta\alpha_{TRef}.$$

We assume further that

$$\Delta\alpha_{Ref} \ll 1, \qquad (28.9)$$

i.e., we are concerned with fairly narrow directivity characteristics for the receiving arrays.

Then, on the basis of (28.6)-(28.8), we obtain

$$R_{12}(\Delta\alpha_0) \approx \exp\left[-(\pi/4)\,(\Delta\alpha_0/\Delta\alpha_{Ref})^2\right]. \qquad (28.10)$$

To estimate the effective angle $\Delta\alpha_{0\,\mathrm{ef}}$ characterizing the reduction in cross-correlation with increasing $\Delta\alpha_0$, we use the integral parameter

$$\Delta\alpha_{0\,\mathrm{ef}} = \int_0^{2\pi} R_{12}(\Delta\alpha_0)\, d(\Delta\alpha_0), \tag{28.11}$$

which is interpretable as the angular cross-correlation interval of reverberation signals arriving from different directions.

Substituting (28.10) into (28.11), and making use of the condition (28.9), we find that

$$\Delta\alpha_{0\,\mathrm{ef}} \approx \Delta\alpha_{\mathrm{Ref}}. \tag{28.12}$$

We see that the angular correlation interval in this case is determined by the effective width of the directivity characteristics of the receiving arrays.

We note that the resultant relations may be used as a basis for investigation of the correlation properties of reverberation signals for various forms of directivity characteristics on the part of the transmitting and receiving acoustic arrays.

If scattering occurs at inhomogeneities distributed throughout the unbounded volume of the ocean medium, it is necessary for analysis of the cross-correlation to bring into the discussion the directivity characteristics $\varphi_{\mathrm{T}}(\alpha, \theta)$, $\varphi_{\mathrm{R1}}(\alpha, \theta)$, and $\varphi_{\mathrm{R2}}(\alpha, \theta)$, which depend on two angular coordinates, as well as the probability density $W(\alpha, \theta)$.

Applying the general relation (7.3) under hypotheses similar to those above, we obtain for the cross-correlation coefficient

$$R_{12}(\Delta\alpha_0, \Delta\theta_0) \approx (1/\Delta\Omega_{\mathrm{TRef}}) \int_{-\pi/2}^{\pi/2} \int_0^{2\pi} \varphi_{\mathrm{T}}^2(\alpha, \theta)\, \varphi_{\mathrm{R}}(\alpha + \Delta\alpha_0/2, \theta + \Delta\theta_0/2) \times$$

$$\varphi_{\mathrm{R}}(\alpha - \Delta\alpha_0/2, \theta - \Delta\theta_0/2)\, d\alpha \cos\theta\, d\theta, \tag{28.13}$$

where the parameter

$$\Omega_{\mathrm{TRef}} = \int_{-\pi/2}^{\pi/2} \int_0^{2\pi} \varphi_{\mathrm{T}}^2(\alpha, \theta)\, \varphi_{\mathrm{R}}^2(\alpha, \theta)\, d\alpha \cos\theta\, d\theta \tag{28.14}$$

characterizes the joint directivity of the transmitting and receiving arrays.

§ 29.   Joint Frequency – Time Correlation

For the description of certain properties of reverberation as a stochastic process, as well as for a number of applications, it is of interest to investigate the joint frequency–time correlation of reverberation signals. The frequency–time correlation makes it possible to carry out additional investigations of the features of sound scattering by various inhomogeneities of the ocean medium, to analyze the scattering model adopted for analysis of the statistical properties of reverberation, and to make appropriate estimates of the joint space–frequency–time correlation functions and the so-called ambiguity functions (see Secs. 12, 39).

Let us consider the problem of determining the frequency–time correlation function.

Let two signals $s_1(t)$ and $s_2(t)$ of identical form be transmitted, with spectra having a mutual frequency separation Ω, so that in complex form these signals may be written

$$\left. \begin{array}{l} s_1(t) = s_0(t) \exp\{j\,[(\omega_0 - \Omega)\,t + \Phi(t)]\}, \\ s_2(t) = s_0(t) \exp\{j\,[\omega_0 t + \Phi(t)]\}. \end{array} \right\} \tag{29.1}$$

Table 29.1. Frequency Correlation Coefficient

No.	Signal envelope form	$r_{\mathrm{r}}(0,\Omega)$
1	Rectangular $1, \|t\| \leqslant \delta/2$	$\dfrac{\sin(\Omega\delta/2)}{\Omega\delta/2}$
2	Bell-shaped $\exp[-(t/t_0)^2], \qquad \delta_{\mathrm{ef}} = \sqrt{\dfrac{\pi}{2}} t_0$	$\exp[-(\Omega\delta_{\mathrm{ef}})^2/4\pi]$
3	Exponential $\exp(-t/t_0), \quad t \geqslant 0, \quad \delta_{\mathrm{ef}} = t_0/2$	$[1+(\Omega\delta_{\mathrm{ef}})^2]^{-1/2}$

The joint frequency–time correlation function $b_{\mathrm{r}}(\tau, \Omega)$, according to its definition (see Sec. 6, Corollary 3) in application to the analysis of the statistical relationship between the reverberation processes $V_1(t, \Omega)$ and $V_2(t, \Omega)$ corresponding to transmission of the signals $s_1(t)$ and $s_2(t)$, is written as the modulus of the product mean of the correlated processes, i.e.,

$$b_{\mathrm{r}}(\tau, \Omega) = |\langle V_1^*(t, \Omega) V_2(t + \tau)\rangle|, \tag{29.2}$$

where the process $V^*(t, \Omega)$ is the complex conjugate of the process $V(t, \Omega)$.

On the basis of Eqs. (6.22), (29.1), and (29.2) we obtain for $b_{\mathrm{r}}(\tau, \Omega)$

$$b_{\mathrm{r}}(\tau, \Omega) = \langle n_1 \rangle \langle a^2 \rangle \left| \int_{-\infty}^{\infty} s_0(t) s_0(t + \tau) \exp\{j[\Omega t + \Phi(t + \tau) - \Phi(t)]\} dt \right|. \tag{29.3}$$

From (29.3) it is possible to obtain the frequency–time correlation coefficient, which turns out to be equal to

$$r_{\mathrm{r}}(\tau, \Omega) = (1/\delta_{\mathrm{ef}}) \left| \int_{-\infty}^{\infty} s_0(t) s_0(t + \tau) \exp\{j[\Omega t + \Phi(t + \tau) - \Phi(t)]\} dt \right|. \tag{29.4}$$

If we allow the time shift in (29.4) to be zero, we arrive at the concept of the reverberation frequency correlation coefficient

$$r_{\mathrm{r}}(0, \Omega) = (1/\delta_{\mathrm{ef}}) \left| \int_{-\infty}^{\infty} s_0^2(t) \exp(j\Omega t) dt \right|. \tag{29.5}$$

It is apparent from this that the frequency correlation of reverberation is determined by the form of the transmitted signal envelope and does not depend on the function $\Phi(t)$ specifying the frequency modulation law.

On the other hand, if we let the frequency shift Ω in (29.4) be equal to zero, $r_{\mathrm{r}}(\tau, 0)$ then turns out to be equal to the envelope of the reverberation autocorrelation coefficient envelope:

$$r_{\mathrm{r}}(\tau, 0) \approx r_{\mathrm{r}}(\tau) \approx (1/\delta_{\mathrm{ef}}) \int_{-\infty}^{\infty} s_0(t) s_0(t + \tau) \cos[\Phi(t + \tau) - \Phi(t)] dt, \tag{29.6}$$

since, for quasi-harmonic transmitted signals,

$$\left| \int_{-\infty}^{\infty} s_0(t)\, s_0(t+\tau) \exp\{j\,[\Phi(t+\tau) - \Phi(t)]\}\, dt \right| \approx \int_{-\infty}^{\infty} s_0(t)\, s_0(t+\tau) \cos\,[\Phi(t+\tau) - \Phi(t)]\, dt.$$

The results of calculations of the reverberation frequency correlation coefficient according to Eq. (29.5) are shown in Table 29.1 for three types of signals with various envelope configurations.

It is an interesting fact that the frequency correlation coefficient of reverberation coincides with the amplitude spectrum of the square of the transmitted-signal envelope. Consequently, it follows directly from Eq. (29.5) on the basis of the convolution theorem that

$$r_{\mathrm{I}}(0, \Omega) = (\delta_{\mathrm{ef}}/2\pi) \left| \int_{-\infty}^{\infty} g_0(\omega)\, g_0(\Omega - \omega)\, d\omega \right|, \tag{29.7}$$

where

$$g_0(\omega) = \int_{-\infty}^{\infty} s_0(t) \exp(-j\omega t)\, dt$$

is the spectrum of the signal envelope. This means, in particular, that the frequency correlation interval of the reverberation cannot be less than the effective width of the spectrum of the transmitted-signal envelope.

§ 30.   Correlation with the Scatterers in Motion

The analysis of the correlation characteristics of reverberation signals has been carried out so far without any regard for the influence of possible motion on the part of the scatterers in the ocean medium or displacement of the acoustic arrays. In a number of cases, however, the motion and its dynamics may exert a significant influence on the statistical properties of reverberation, particularly on the correlative, spectral, and other characteristics, this influence being most clearly pronounced in the transmission of fairly narrow-band signals, i.e., signals of long duration.

Allowance for the factors just named permits us to describe the true picture of sound scattering in the observation, specifically, of surface reverberation when the scatterers are moving under the influence of ocean surface waves or volume reverberation in the observation of scattering, for example, by schools of fish, accumulations of various types of marine life, or other moving scatterers lumped together or otherwise localized in space. With this in mind, in the present and succeeding sections we examine the correlation characteristics of reverberation observed under conditions of moving scatterers and acoustic arrays.

We proceed first with an analysis of scatterer motion and its influence on the reverberation autocorrelation function. We employ the following model for motion of the scatterers:

The motion of the scatterers in the ocean follows independent trajectories.

During a period equal to the effective duration of the transmitted signals, the velocities of the scatterers are constant in magnitude and direction.

The relative variations of the middle frequency $\Delta\omega_i$ of the spectrum of the elementary scattered signals are relatively small, so that

$$\Delta\omega_i/\omega_0 \ll 1.$$

In order to determine the reverberation autocorrelation function in the case in question, it is necessary to have specified the probability density $W(\Delta\omega)$ of the stochastic variable $\Delta\omega_i$ characterizing the frequency shift of the elementary scattered signals and to apply a generalization of the two-dimensional theorem of superposition of stochastic perturbations (see Sec. 7). The stochastic parameter $\Delta\omega_i$ is related to the velocity c_i of the i-th scatterer by the classical relation expressing the Doppler effect:

$$\Delta\omega_i = (2v_i/c)\,\omega_0,$$

where c is the velocity of sound propagation.

The reverberation autocorrelation function $B_{rs}(\tau)$ is determined in the following form on the basis of the general relation (7.3):

$$B_{rs}(\tau) = \langle n_1 \rangle \langle a^2 \rangle \int\limits_{-\infty}^{\infty} W(\Delta\omega) \int\limits_{-\infty}^{\infty} s(t, \Delta\omega)\, s(t + \tau, \Delta\omega)\, dt\, d(\Delta\omega), \tag{30.1}$$

where, in the given instance,

$$s(t, \Delta\omega) = s_0(t) \cos\left[(\omega_0 + \Delta\omega)\, t + \Phi(t)\right]. \tag{30.2}$$

Substituting s(t, $\Delta\omega$) according to (30.2) into (30.1), we obtain

$$B_{rs}(\tau) = \langle n_1 \rangle \langle a^2 \rangle \int\limits_{-\infty}^{\infty} W(\Delta\omega) \int\limits_{-\infty}^{\infty} s_0(t)\, s_0(t + \tau) \cos \times$$
$$\times \left[(\omega_0 + \Delta\omega)\, t + \Phi(t)\right] \cos\left[(\omega_0 + \Delta\omega)(t + \tau) + \Phi(t + \tau)\right] dt\, d(\Delta\omega). \tag{30.3}$$

We next take into account the quasi-harmonicity of the transmitted signals and assume that the probability density $W(\Delta\omega)$ is an even function with respect to $\Delta\omega = 0$. Then, after some straightforward trigonometric manipulations of the integrand in (30.3), we have for $B_{rs}(\tau)$:

$$B_{rs}(\tau) \approx \langle n_1 \rangle \langle a^2 \rangle \left\{ 2 \int\limits_{-\infty}^{\infty} W(\Delta\omega) \cos(\Delta\omega\tau)\, d(\Delta\omega) \times \right.$$
$$\left. \times \int\limits_{-\infty}^{\infty} s_0(t)\, s_0(t + \tau) \cos\left[\Phi(t + \tau) - \Phi(t)\right] dt \right\} \cos\omega_0\tau. \tag{30.4}$$

It is a fairly simple matter in Eq. (30.4) to perceive the multiplicativeness of the parameters characterizing the statistical properties of the motion of the scatterers and the form of the transmitted signals. The first parameter, which is equal to

$$\Theta(\tau) = 2 \int\limits_{0}^{\infty} W(\Delta\omega) \cos(\Delta\omega\tau)\, d(\Delta\omega), \tag{30.5}$$

is nothing other than the characteristic function of the probability density $W(\Delta\omega)$.

The second parameter,

$$B_r(\tau) \approx \langle n_1 \rangle \langle a^2 \rangle \left\{ \int\limits_{-\infty}^{\infty} s_0(t)\, s_0(t + \tau) \cos\left[\Phi(t + \tau) - \Phi(t)\right] dt \right\} \cos\omega_0\tau, \tag{30.6}$$

represents the reverberation autocorrelation function, corresponding to zero motion of the scatterers, when their velocity distribution is described by a delta function, i.e.,

$$W(\Delta\omega) = \delta(\Delta\omega). \tag{30.7}$$

Consequently, Eqs. (30.4)-(30.6) imply

$$R_{rs}(\tau) = \Theta(\tau) \, B_r(\tau). \tag{30.8}$$

Also self-evident is the expression for the correlation coefficient

$$R_{rs}(\tau) = \Theta(\tau) r_r(\tau) \cos \omega_0 \tau. \tag{30.9}$$

If the distribution $W(\Delta\omega)$ is symmetrical with respect to ω_s, it becomes necessary to substitute the following into the general relation (30.1) instead of (30.2):

$$s(t) = s_0(t) \, \cos[(\omega_0 + \omega_s + \Delta\omega)t + \Phi(t)],$$

and we arrive at a relation similar to (30.9), except that it contains the factor $\cos[(\omega_0 + \omega_s)\tau]$ in place of $\cos \omega_0 \tau$, i.e.,

$$R_{rs}(\tau) = \Theta(\tau) \, r_r(\tau) \cos[(\omega_0 + \omega_s)\tau]. \tag{30.10}$$

Now the characteristic function $\theta(\tau)$ is found from the centered distribution law $W(\Delta\omega)$, $\langle\Delta\omega\rangle = 0$.

In the special case when the scatterers are moving at a constant relative velocity v_s, so that $\omega_s = (2v_s/c)\omega_0$, we have, by virtue of (30.5) and (30.7), $\Theta(\tau) = 1$, and Eq. (30.10) implies

$$R_{rs}(\tau) = r_r(\tau) \cos[(\omega_0 + \omega_s)\tau]. \tag{30.11}$$

It is also interesting to look at the case when the motion of the scatterers is such that they may be divided into several groups, each of which has its own velocity distribution law. We then write

$$W(\Delta\omega) = \sum_{i=1}^{N} P_i W_i(\omega_{si} + \Delta\omega), \tag{30.12}$$

where $W_i(\omega_{si} + \Delta\omega)$ is the probability density of the frequency shifts of the i-th set of elementary scattered signals, the P_i are weighting factors, the sum of which due to the normalization condition

$$\int_{-\infty}^{\infty} W(\Delta\omega) \, d(\Delta\omega) = 1, \quad \int_{-\infty}^{\infty} W_i(\omega_{si} + \Delta\omega) \, d(\Delta\omega) = 1$$

is equal to unity, i.e.,

$$\sum_{i=1}^{N} P_i = 1. \tag{30.13}$$

Making use of the general relation (30.1) and substituting (30.12) therein, we obtain for $B_{rs}(\tau)$,

$$B_{rs}(\tau) = \langle n_1 \rangle \langle a^2 \rangle \sum_{i=1}^{N} P_i \int_{-\infty}^{\infty} W_i(\omega_{si} - \Delta\omega) \int_{-\infty}^{\infty} s(t) s(t+\tau) \, dt \, d(\Delta\omega). \tag{30.14}$$

We assume, as before, that the distributions $W_i(\omega_{si} + \Delta\omega)$ are symmetrical with respect to ω_{si}; consequently, every term of the sum (30.14) may be reduced to the form (30.10). Then, clearly, the reverberation correlation coefficient is determined in the form

$$R_{rs}(\tau) \approx r_r(\tau) \sum_{i=1}^{N} P_i \Theta_i(\tau) \cos[(\omega + \omega_{pi})\tau], \tag{30.15}$$

where

$$\Theta_i(\tau) = 2 \int_0^\infty W_i(\Delta\omega) \cos(\Delta\omega\tau) d(\Delta\omega), i=1,2,\ldots,N, \tag{30.16}$$

are the characteristic functions of the centered distributions $W_i(\Delta\omega)$.

When for any i

$$W_i(\Delta\omega) = \delta(\Delta\omega),$$

Eq. (30.15) acquires the simplest form:

$$R_{rs}(\tau) \approx r_r(\tau) \sum_{i=1}^{N} P_i \cos[(\omega_0 + \omega_{si})\tau]. \tag{30.17}$$

Specifying the values of the parameters P_i, ω_{si}, the form of the distributions $W_i(\Delta\omega)$, and the form of the transmitted signal, the expressions derived above may be used to calculate the specific form of the autocorrelation characteristics of reverberation with regard for the influence of the motion of the scatterers.

We consider as an example the case when the motion of the scatterers is specified by the characteristics

$$\left.\begin{array}{ll} N = 2, & P_1 = P, \ P_2 = 1 - P, \\ \omega_{s_1} = 0, & W_1(\Delta\omega) = \delta(\Delta\omega), \\ \omega_{s_2} = \Omega, & W_2(\Delta\omega) = (1/\sqrt{2\pi}\,\sigma_\Omega) \exp(-\Delta\omega^2/2\sigma_\Omega), \end{array}\right\} \tag{30.18}$$

and the envelope of the transmitted pulse, which has a sinusoidal carrier, has the squared exponential form (10.8).

Then, making use of the data in Table 22.1 (row 2), the general relation (30.15), and Eq. (30.16) with the conditions (30.18) and (10.8), we find the reverberation autocorrelation coefficient:

$$B_{rs}(\tau) \approx P \exp[-(\pi/4)(\tau/\delta_{ef})^2] \cos\omega_0\tau +$$

$$+ (1 - P) \exp[-(\pi/4)(\tau/\delta_{ef})^2 - (\tau\sigma_\Omega)^2/2] \cos(\omega_0 + \Omega)\tau. \tag{30.19}$$

Consequently, it follows from Eq. (30.10) that the reverberation correlation coefficient calculated with regard for motion of the scatterers can differ substantially from the correlation coefficient

$$R_r(\tau) = \exp[-(\pi/4)(\tau/\delta_{ef})^2] \cos\omega_0\tau,$$

which corresponds to stationary scatterers, this difference being more appreciable the larger the values of the parameters $1 - P$, σ_Ω, and Ω.

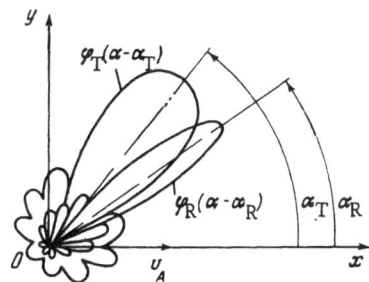

Fig. 44. Diagram for the correlation analysis of reverberation with displacement of the acoustic arrays.

§31. Correlation with the Acoustic Arrays in Motion

It was shown in the investigation of the influence of scatterer motion on the correlation characteristics of reverberation that these characteristics can vary within wide limits due to the Doppler effect, depending on the distribution of the relative scatterer velocities. Displacement of the acoustic arrays, as often occurs in the observation of reverberation signals under real circumstances, can lead to similar results. Such an effect is reasonable in view of the relativity properties of the motion, whence follows the common ground on which the physical models of sound scattering in both cases is based.

The analysis of the correlation of reverberation with motion of the acoustic arrays is carried out for the example of sound scattering by inhomogeneities concentrated in an unbounded thin layer. For scattering by inhomogeneities distributed throughout the entire volume of space or in some part of it, the correlation characteristics of reverberation are investigated similarly to a large degree, so that the results of the present section may be readily generalized to the case of three-dimensional scattering.

For example, in the case of a thin unbounded scattering layer, let a transmitter and receiver with normalized directivity characteristics $\varphi_T(\alpha)$ and $\varphi_R(\alpha)$ be situated at one point, and let them move with a constant velocity v_A (Fig. 44).

If we specify the probability density $W(\alpha)$, we have for $B_{rA}(\tau)$ on the basis of the general expression (7.3)

$$B_{rA}(\tau) = \langle n_1 \rangle \int_0^{2\pi} \langle a^2(\alpha) \rangle\, W(\alpha) \int_{-\infty}^{\infty} s(t, \alpha)\, s(t + t, \alpha)\, dt\, d\alpha. \tag{31.1}$$

The quantity $\langle a^2(\alpha) \rangle$ depends on the angle α and is related to the directivity characteristics of the transmitting and receiving arrays by the equation

$$\langle a^2(\alpha) \rangle = \langle a^2 \rangle\, \varphi_T^2(\alpha - \alpha_T)\, \varphi_R^2(\alpha - \alpha_R). \tag{31.2}$$

We note that the angles α_T and α_R in Eq. (31.2) are measured from the line of motion of the arrays and give the axial directions of the normalized characteristics $\varphi_T(\alpha)$ and $\varphi_R(\alpha)$ (see Fig. 44). This means that

$$\varphi_T(0) = 1, \quad \varphi_R(0) = 1.$$

The function s(t) is determined by the form of the transmitted signals, the velocity v_A of the acoustic arrays, and the angle α. It may be given approximately as follows:

$$s(t, \alpha) \approx s_0(t) \cos\,[\omega_0\,(1 + 2v_A \cos\,\alpha/c)t + \Phi(t)], \tag{31.3}$$

where, as in Sec. 30, it is assumed that

$$v_A/c \ll 1. \tag{31.4}$$

If this inequality is weakly satisfied, it is then required in Eq. (31.3) to take account of the variation in form of the modulating functions $s_0(t)$ and $\Phi(t)$.

Substituting Eqs. (31.2) and (31.3) into (31.1), we obtain for the reverberation autocorrelation function

$$B_{rA}(\tau) \approx \langle n_1 \rangle \langle a^2 \rangle \int\limits_0^{2\pi} \varphi_T^2(\alpha - \alpha_T) \varphi_R^2(\alpha - \alpha_R) W(\alpha) \int\limits_{-\infty}^{\infty} s_0(t) s_0(t+\tau) \times$$
$$\times \cos[\omega_0(1 + 2v_A \cos\alpha/c)\, t + \Phi(t)] \times$$
$$\times \cos[\omega_0(1 + 2v_A \cos\alpha/c)(t+\tau) + \Phi(t+\tau)]\, dt\, d\tau\, d\alpha. \qquad (31.5)$$

If then we regard the condition of quasi-harmonicity of the transmitted signals, as well as the inequality (31.4), we go from (31.5) to the following relation:

$$B_{rA}(\tau) \approx (\langle n_1 \rangle \langle a^2 \rangle \, \theta_{ef}/2) \, r_r(\tau) \int\limits_0^{2\pi} \varphi_T^2(\alpha - \alpha_T) \varphi_R^2(\alpha - \alpha_R) W(\alpha) \times$$
$$\times \cos[\omega_0(1 + 2v_A \cos\alpha/c)\,\tau]\, d\alpha, \qquad (31.6)$$

where $r_r(\tau)$ is the reverberation correlation coefficient envelope (21.7), calculated on the assumption of nonmoving acoustic arrays, when $v_A = 0$.

Proceeding on the basis of Eq. (31.6), the reverberation correlation coefficient turns out to be equal to

$$R_{rA}(\tau) \approx [r_r(\tau)/\Delta\alpha_{TR}] \int\limits_1^{2\pi} \varphi_T^2(\alpha - \alpha_T) \varphi_R^2(\alpha - \alpha_R) W(\alpha) \cos[\omega_0(1 + 2v_A \cos\alpha/c)\,\tau]\, d\alpha, \qquad (31.7)$$

where the parameter

$$\Delta\alpha_{TR} = \int\limits_0^{2\pi} \varphi_T^2(\alpha - \alpha_T) \varphi_R^2(\alpha - \alpha_R) W(\alpha)\, d\alpha \qquad (31.8)$$

characterizes the distribution of the scatterers in space and the directionality of the arrays.

Equation (31.7) makes it possible to calculate the correlation characteristics of reverberation for various directivity patterns of the acoustic arrays, their velocities of motion, and distributions of the scatterers in space.

Let us consider some examples.

If the acoustic arrays are nondirectional and the scatterers are uniformly distributed throughout a layer, i.e., if

$$\left.\begin{array}{ll} \varphi_T(\alpha) = 1, & \varphi_R(\alpha) = 1, \\ W(\alpha) = 1/2\pi, & 0 \leqslant \alpha < 2\pi, \end{array}\right\} \qquad (31.9)$$

then we have for $R_{rA}(\tau)$, on the basis of (31.7) and (31.9), clearly,

$$R_{rA}(\tau) \approx r_r(\tau)\, J_0(2kv_A\tau) \cos\omega_0\tau, \qquad (31.10)$$

where $J_0(X)$ is the Bessel function of a real argument (27.18), k is the wave number.

Another example is the following:

$$\varphi_T(\alpha) = 1; \qquad \varphi_R(\alpha) = 1 \quad \text{for} \quad \alpha_R - \Delta\alpha_R/2 \leqslant \alpha < \alpha_R + \Delta\alpha_R/2;$$
$$W(\alpha) = 1/2\pi \quad \text{for} \quad 0 \leqslant \alpha < 2\pi. \qquad (31.11)$$

This case is equivalent to nondirectional transmission and reception by an acoustic array with a so-called sectoral directivity characteristic. Assuming that $\Delta\alpha_R \ll 1$, we obtain for the correlation coefficient on the basis of (31.7) and (31.11)

$$R_{rA}(\tau) \approx r_r(\tau) \cos\left[\omega_0\left(1 + 2v_A \cos\varkappa_R/c\right)\tau\right]. \tag{31.12}$$

It is interesting to note that a relation of the type (31.12) may be derived by letting

$$W(\alpha) = \delta(\alpha - \alpha_R)$$

and assuming, for example, that the acoustic arrays are nondirectional.

In the general case of an arbitrary scatterer distribution in space and various forms of $\varphi_T(\alpha)$ and $\varphi_R(\alpha)$, as is readily inferred from Eqs. (31.10) and (31.12), it is possible to have simultaneous variation of the correlation coefficient envelope and shifting of its oscillation frequency.

CHAPTER V

SPECTRAL ANALYSIS OF REVERBERATION

§ 32. Preliminary Remarks

In the preceding chapter we investigated the correlation characteristics of reverberation. We also studied its spectral properties.

The statistical spectrum or, as it is sometimes called, the <u>power spectrum</u> $G(\omega)$ of stochastic processes and its correlation function $B(\tau)$ are related in one-to-one fashion by a pair of Fourier transforms:

$$G(\omega) = 2 \int_{-\infty}^{\infty} B(\tau) \exp(-j\omega\tau) d\tau, \tag{32.1}$$

$$B(\tau) = (1/4\pi) \int_{-\infty}^{\infty} G(\omega) \exp(j\omega\tau) d\omega. \tag{32.2}$$

These relations reflect the essential content of the Wiener−Khinchin theorem, the central theorem of harmonic analysis, which is valid primarily for stationary stochastic processes. The spectrum and correlation function of stationary processes are even functions of their arguments, so that (32.1) and (32.2) may be rewritten in real form:

$$G(\omega) = 4 \int_{0}^{\infty} B(\tau) \cos \omega\tau \, d\tau, \tag{32.3}$$

$$B(\tau) = (1/2\pi) \int_{0}^{\infty} G(\omega) \cos \omega\tau \, d\omega. \tag{32.4}$$

Hence, it follows that, given a known reverberation correlation function, it is always possible in principle to find its energy spectrum. Such a one-to-one connection, considering that the correlation characteristics of reverberation are based on the two-dimensional theorems of superposition of stochastic perturbations, conveys the impression that no further analysis of the frequency spectra of reverberation is needed. However, the investigation of these spectra has an unquestionable intrigue all its own, for the following reasons.

First, spectral analysis makes it possible in several cases to describe particular features of reverberation observed under varying conditions more transparently than does correlation analysis. The opportunity is provided here for reliable experimental verification of the sound-scattering model at inhomogeneities of various types, as well as an explanation of the limits of validity of the assumptions made in calculating the statistical characteristics of reverberation. The reverberation spectra in this case can be calculated directly in some cases from the characteristics of the transmitted signals without preliminary correlation analysis.

Second, inasmuch as spectral analysis is simpler than correlation analysis from the instrumental point of view, * the comparison of analytical and experimental data is often effected more conveniently by comparing spectra, especially since standard analyzers guarantee a much higher resolving power and more acceptable speed of analysis.

Third, many problems in the processing of reverberation signals, particularly by means of linear systems, are solved more simply with the application of spectral analytic methods, which are found in wide use because of their ready visualization.

In the present chapter we investigate two types of reverberation spectra, namely, the spectra of the actual reverberation and the spectra of its envelope fluctuations.

§ 33.   General Relations for the Reverberation Spectrum

Here we find expressions for the energy spectrum of reverberation under the condition that the form of the elementary scattered signals follows that of the transmitted pulses. This assumption is justified in practice, for example, in the case of the transmission of short or wide-band signals, when the influence of the motion of the scatterers and acoustic arrays may be neglected in the first approximation. The spectral characteristics of reverberation are easily determined by using the appropriate expressions derived for its autocorrelation functions in the preceding chapter.

If the transmitted signals are described by functions s(t) of the determinate type, the reverberation correlation function is determined by the relation

$$B_r(\tau) = \langle n_1 \rangle \langle a^2 \rangle \int_{-\infty}^{\infty} s(t)\, s(t + \tau)\, dt. \tag{33.1}$$

We substitute Eq. (33.1) into (32.1). Then, for the statistical spectrum of reverberation $G_r(\omega)$, we obtain

$$G_r(\omega) = 2 \langle n_1 \rangle \langle a^2 \rangle \int_{-\infty}^{\infty} \int_{-\infty}^{\infty} s(t)\, s(t + \tau) \exp(-j\omega\tau)\, d\tau\, dt. \tag{33.2}$$

After integration over the variable τ, with consideration for the fact that

$$g_s(\omega) = \int_{-\infty}^{\infty} s(t) \exp(-j\omega t)\, dt \tag{33.3}$$

represents the amplitude spectrum of the transmitted signal, we rewrite (33.2) in the form

$$G_r(\omega) = 2 \langle n_1 \rangle \langle a^2 \rangle\, g_s(\omega) \int_{-\infty}^{\infty} s(t) \exp(j\omega t)\, dt. \tag{33.4}$$

---

*We stress the fact that we are speaking here only of the technical advantage of spectral analysis, not its conceptual advantage; from the viewpoint of the general notions regarding the correctness of using particular methods for measuring the statistical characteristics of stochastic process, the situation, unfortunately, is reversed.

The integral on the right-hand side of (33.4) represents the complex conjugate spectrum of the transmitted signal; we denote it by

$$g_s^{\bullet}(\omega) = \int_{-\infty}^{\infty} s(t) \exp(j\omega t)\, dt, \quad g_s^{\bullet}(\omega) = g_s(-\omega). \tag{33.5}$$

We next introduce the squared modulus of the amplitude spectrum:

$$|g_s(\omega)|^2 = g_s(\omega)\, g_s^{\bullet}(\omega). \tag{33.6}$$

Then, on the basis of (33.4)-(33.6), we arrive at the final expression for the statistical spectrum of reverberation:

$$G_{\mathbf{r}}(\omega) = 2 \langle n_1 \rangle \langle a^2 \rangle |g_s(\omega)|^2. \tag{33.7}$$

Consequently, the statistical spectrum of reverberation in the transmission of determinate-type signals is proportional to the square of their amplitude spectrum.

In addition to (33.7), it is sometimes helpful to use the expression for the normalized statistical spectrum:

$$G_{\mathbf{rn}}(\omega) = g_{\mathbf{rn}}^2(\omega) = \frac{|g_s(\omega)|^2}{|g_s(\omega)|_{\max}^2}, \tag{33.8}$$

which is obtained by normalization of $G_{\mathbf{r}}(\omega)$ to its maximum value.

In the transmission of quasi-harmonic signals the peak values of the spectrum $g_S(\omega)$ fall at the frequencies $\omega = \omega_0$ and $\omega = -\omega_0$. This permits a determination of the normalized rms spectrum in the form of the ratio

$$g_{\mathbf{rn}}(\omega) = \frac{|g_s(\omega)|}{|g_s(\omega_0)|}. \tag{33.9}$$

We next derive an expression for the statistical spectrum of reverberation for the transmission of noisy signals. In this case (see Sec. 23)

$$B_{\mathbf{r}}(\tau) = \langle n_1 \rangle \langle a^2 \rangle R_N(\tau) \int_{-\infty}^{\infty} s_0(t)\, s_0(t+\tau)\, dt, \tag{33.10}$$

where $R_N(\tau)$ is the autocorrelation coefficient of the noise carrier.

Applying the general formula (32.1) to (33.10), we obtain

$$G_{\mathbf{r}}(\omega) = 2 \langle n_1 \rangle \langle \alpha^2 \rangle \int_{-\infty}^{\infty} \int_{-\infty}^{\infty} R_N(\tau)\, s_0(t) s_0(t+\tau) \exp(-j\omega\tau)\, d\tau dt. \tag{33.11}$$

Letting

$$r_{\mathbf{r}}(\tau) = (1/\delta_{ef}) \int_{-\infty}^{\infty} s_0(t)\, s_0(t+\tau)\, dt \tag{33.12}$$

denote a function having the sense of the autocorrelation coefficient of the normalized determinate envelope of the pulse, we go from (33.11) to the relation

$$G_r(\omega) = 2 \langle n_1 \rangle \langle a^2 \rangle \delta_{ef} \int_{-\infty}^{\infty} R_N(\tau)\, r_r(\tau) \exp(-j\omega\tau)\, d\tau. \tag{33.13}$$

We then denote by

$$\left.\begin{aligned} G_N(\omega) &= 2 \int_{-\infty}^{\infty} R_N(\tau) \exp(-j\omega\tau)\, d\tau, \\ G_0(\omega) &= 2 \int_{-\infty}^{\infty} r_r(\tau) \exp(-j\omega\tau)\, d\tau \end{aligned}\right\} \tag{33.14}$$

the statistical spectra of the noise carrier and the determinate envelope of the transmitted-signal autocorrelation coefficient. Then, applying the convolution theorem to (33.13), we obtain

$$G_r(\omega) = (\langle n_1 \rangle \langle a^2 \rangle \delta_{ef}/4\pi) \int_{-\infty}^{\infty} G_N(v)\, G_0(\omega - v)\, dv. \tag{33.15}$$

Consequently, <u>the statistical spectrum of reverberation in the transmission of noisy signals is proportional to the convolution of the energy spectra of the noise carrier and the determinate envelope.</u>

It is interesting to note the limiting case when the noise carrier degenerates to harmonic, i.e.,

$$R_N(\tau) = \cos \omega_0 \tau. \tag{33.16}$$

Here we obtain from (33.14) and (33.16) the fact that

$$G_N(\omega) = 2\pi\, [\delta(\omega - \omega_0) + \delta(\omega + \omega_0)].$$

Substituting the value of $G_N(\omega)$ into the general expression (33.15), we find

$$G_r(\omega) = (\langle n_1 \rangle \langle a^2 \rangle \delta_{ef}/2)\, [G_0(\omega - \omega_0) + G_0(\omega + \omega_0)]. \tag{33.17}$$

We have arrived at a logical result, which indicates the fact that in the transmission of signals with a sinusoidal carrier the reverberation spectrum consists of two components with central frequencies corresponding to $\omega = \omega_0$ and $\omega = -\omega_0$, and that the shape of the spectrum is uniquely determined by the envelope of the transmitted signal.

§ 34. Reverberation Spectra in the Transmission of Various Types of Signals

In this section we are concerned primarily with experimental data on the spectral composition of reverberation signals and the analysis of their consistency with the analytical characteristics. The results of the measurements refer chiefly to surface and bottom reverberation and, in a few cases, to their combination.

The conditions of the measurements were usually characterized by a uniform temperature distribution with respect to depth (winter period), so that the effects of sound refraction at distances of a few kilometers was scarcely felt. In some instances the measurements were performed in the spring, when there was a strong negative temperature gradient in the surface layer.

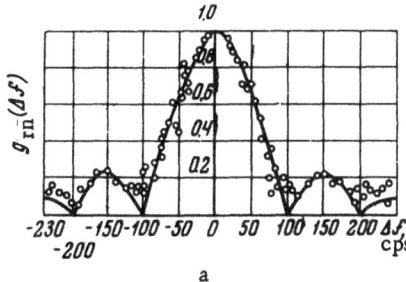

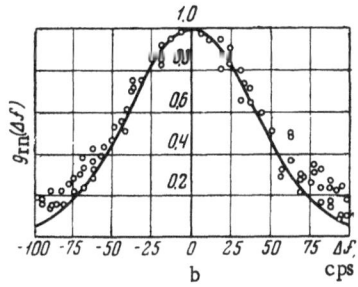

Fig. 45. Analytical (curves) and measured (circles) normalized mean-square reverberation spectra. Shape of the transmitted pulse envelope. a) Rectangular; b) bell-shaped. $f_0 = 15$ kc, $\delta = 10$ msec.

The results of several series of measurements of the mean-square reverberation spectra for the transmission of various types of pulses are discussed below.

In one series of measurements the reverberation spectra were obtained for the transmission of pulses with a sinusoidal carrier and two types of envelopes: rectangular and bell-shaped.

In conducting this series of measurements, the conditions of sound propagation were characterized by a uniform temperature distribution with respect to depth, and at the point of reception mainly surface reverberation was detected; the surface wave state was two or three points on the sea-height scale.

The measured reverberation spectra are shown in Fig. 45 in comparison with the theoretical curves. The calculated normalized mean-square spectra were found for the case in question by substituting the data of Table 11.1 into Eq. (33.9). It was found that in the transmission of rectangular pulses

$$g_{rn}(\Delta\omega) = \left| \frac{\sin(\Delta\omega\delta/2)}{\Delta\omega\delta/2} \right|, \qquad \Delta\omega = \omega - \omega_0; \qquad (34.1)$$

while in the transmission of bell-shaped pulses

$$g_{rn}(\Delta\omega) = \exp\left[-(\Delta\omega\delta_{ef})^2/2\pi \right]. \qquad (34.2)$$

The procedure for the spectral measurements was set up to provide for processing of the stationarized reverberation time intervals, including 10 to 30 independent readings of the process and subsequent averaging of the results of a partial analysis over 25-30 reverberation events.

It follows from Fig. 45 that the agreement between the measured and calculated reverberation spectra is fully satisfactory. A certain broadening observed in the spectrum for transmission of bell-shaped pulses is apparently due to the influence of scatterer motion in the surface layer or inadequate precision in the analysis of the small relative levels of the side frequency components.

The experimental investigations of the reverberation spectra in the transmission of long-duration pulses, lasting, for example, hundreds of milliseconds, showed that the relative broadening of the spectra increases in general with the duration of the signals and with the carrier frequency. These effects are consistent with the notions of scatterer motion and its influence on the statistical characteristics of reverberation signals.

On the other hand, with a decrease in the duration of the transmitted pulses, better agreement is observed between the calculated and experimental reverberation spectra.

Similar results were obtained in [37], in which the spectra of bottom reverberation for pulses with a rectangular envelope and various durations were compared. In this article good

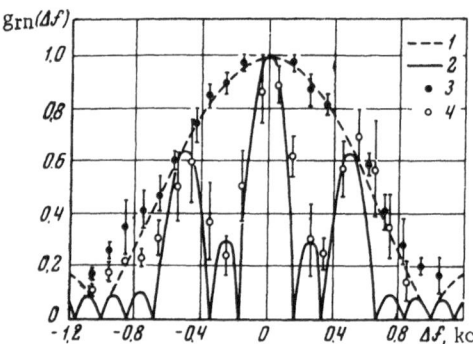

Fig. 46. Normalized spectra for the transmission of a rectangular-pulse train. 1) Analytical amplitude spectrum of the individual pulse; 2) analytical mean-square reverberation spectrum; 3) measured individual-pulse spectrum; 4) measured mean-square reverberation spectrum. $N = 3$, δ_1 = 1 msec, T_0 = 2 sec, f_0 = 26 kc.

agreement was obtained between the measured mean-square reverberation spectra and the moduli of the amplitude spectra of pulses for δ = 1, 0.5, and 0.2 msec for f_0 = 100 kc.

We also investigated the reverberation spectra for the transmission of a train of rectangular pulses. If the train consists of N identical pulses of duration δ_1 following one another with a repetition period equal to T_0, we have for the normalized mean-square reverberation spectrum $g_{rn}(\Delta\omega)$, on the basis of (10.19) and (33.9),

$$g_{rn}(\Delta\omega) = \left| \frac{\sin(\Delta\omega\delta_1/2)}{\Delta\omega\delta_1/2} \right| \left| \frac{\sin(\Delta\omega NT_0/2)}{N\sin(\Delta\omega T_0/2)} \right|. \quad (34.3)$$

We note that for pulses $s_1(t)$ of arbitrary shape in a train, the following more general relation is valid for $g_{rn}(\Delta\omega)$:

$$g_{rn}(\Delta\omega) = |g_{s1}(\Delta\omega)| \left| \frac{\sin(\Delta\omega NT_0/2)}{N\sin(\Delta\omega T_0/2)} \right|, \quad (34.4)$$

where $g_{s1}(\Delta\omega)$ is the normalized amplitude spectrum of the individual pulse in the train.

It follows from (34.4) that the mean-square reverberation spectrum in this case is significantly nonuniform, having peaks at frequencies corresponding to $\Delta\omega T_0/2 = 0, 1, 2, \ldots$, where the envelope of the principal peaks of the spectral density coincides in shape with the amplitude spectrum of the individual pulse.

The experimental mean-square reverberation spectrum for the transmission of a train of three pulses is shown in Fig. 46 with the analytical spectrum, constructed in accord with the relation (34.3). Also shown in this figure for comparison are the measured and theoretical spectra for the individual pulses of the train. Combined surface and bottom reverberation were subjected to spectral analysis in this case, under the following conditions: surface state of two points on the sea-height scale at the crest; large pebbles for the bottom.

It is evident from a comparison of the spectra that in general the results of the measurements coincide with the theoretical dependences except in regions with low relative levels on the part of the spectral components, where the insufficient resolving power of the analyzer caused values that were too high for the spectral density.

We note that the width of each peak of the reverberation spectrum in the transmission of a pulse train turns out to be inversely proportional to the number of pulses N in the train, while the level of the mean-square reverberation spectrum is proportional to $N^{1.5}$. The latter dependence becomes clear when we consider that the mean-square value of the reverberation increases due to the increasing effective duration of the transmitted signal in proportion to $N^{0.5}$, and that the amplitude spectral density of the pulse itself is proportional to N. Near the fundamental peaks the normalized level of the spectral components $g_{rn}(\Delta\omega)$ is well described by the formula

$$g_{rn}(\Delta\omega) \approx \left| \frac{\sin(\Delta\omega NT_0/2)}{\Delta\omega NT_0/2} \right|, \quad (34.5)$$

which corresponds to the spectrum of a rectangular pulse with duration NT_0.

We next investigate the reverberation spectra for the transmission of a noisy signal.

In this case, the statistical reverberation spectrum may be calculated either by means of the general relation (33.15) or by application of (32.3) to the reverberation correlation function, which is expressed in the form (33.10) for the transmission of noisy signals.

We employ the second method of calculation.

If the transmitted signal represents a segment of stationary noise of duration δ, then

$$G_r(\omega) = 4 \langle n_1 \rangle \langle a^2 \rangle \int_0^\delta R_N(\tau) (1 - \tau/\delta) \cos \omega\tau d\tau. \tag{34.6}$$

For quasi-harmonic noise the autocorrelation coefficient is written in the form

$$R_N(\tau) \approx r_N(\tau) \cos \omega_0\tau.$$

Now, proceeding from (34.6) to the normalized mean-square reverberation spectrum, we obtain

$$g_{rn}(\Delta\omega) = \left[\frac{\int_0^\delta r_N(\tau)(1 - \tau/\delta) \cos(\Delta\omega\tau)\,d\tau}{\int_0^\delta r_N(\tau)(1 - \tau/\delta)\,d\tau} \right]^{1/2}, \quad \Delta\omega = \omega - \omega_0. \tag{34.7}$$

It is apparent from (34.7) that if the correlation interval of the noise carrier is significantly less than the pulse duration,

$$g_{rn}(\Delta\omega) \approx \left[\frac{\int_0^\infty r_N(\tau) \cos(\Delta\omega\tau)\,d\tau}{\int_0^\infty r_N(\tau)\,d\tau} \right]^{1/2}, \tag{34.8}$$

i.e., the reverberation spectrum is uniquely determined by the spectrum of the noise carrier. In the other extreme case we arrive at a relation similar to (34.1), corresponding to the transmission of pulses with a sinusoidal carrier and rectangular envelope.

In the intermediate cases the shape of the reverberation spectrum may be determined both by the signal duration and by the form of the correlation function for its carrier.

As an example, we consider the reverberation spectrum associated with the transmission of noisy pulses, the carrier of which is shaped by a single resonance loop. In this case (see Table 23.1),

$$r_N(\tau) \approx \exp(-\Delta\omega_N |\tau|), \tag{34.9}$$

where $\Delta\omega_N$ is the filter transmission band, referred to the 0.7 level. Substitution of (34.9) into (34.7) and integration lead to the following results:

$$g(\Delta\omega) \approx \frac{k}{k^2 + p^2} \left\{ \frac{\exp(-k)[(k^2 - p^2)\cos p - 2pk\sin p] + k(k^2 + p^2) - (k^2 - p^2)}{\exp(-k) + k - 1} \right\}^{1/2}, \tag{34.10}$$

where

$$k = \Delta\omega_N\delta, \quad p = \Delta\omega\delta.$$

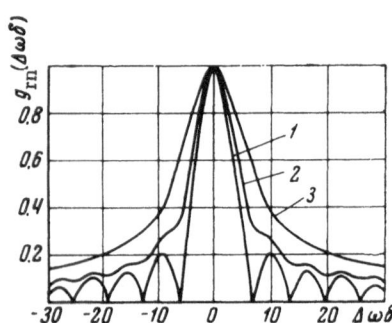

Fig. 47. Normalized mean-square reverbera-tion spectra for the transmission of noise pulses. 1) $\Delta\omega_N\delta = 0$; 2) $\Delta\omega_N\delta = 1$; 3) $\Delta\omega_N\delta = 3$.

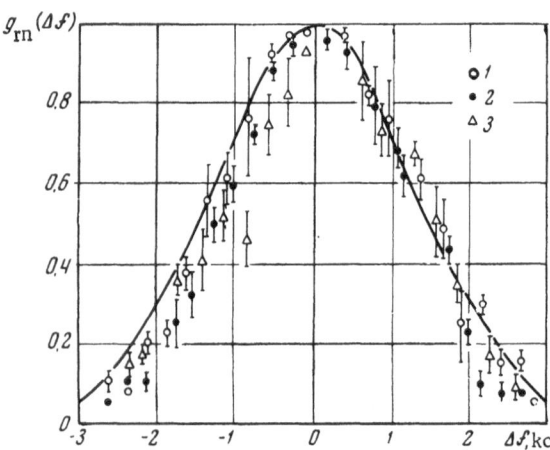

Fig. 48. Measured reverberation spectra for the transmission of noise pulses. 1) $\delta = 3$ msec; 2) $\delta = 10$ msec; 3) $\delta = 20$ msec. The curve represents the normalized frequency characteristic of a frequency with a passband $\Delta F_N = 2.1$ kc.

Table 34.1. Width of the Reverberation Spectrum

	ΔF_r, kc						
Exp. No.	δ , msec						
	0.3	1.0	1.0	5	10	20	30
I	3.5	2.8	—	—	2.1	—	2.1
II	3.2	—	2.6	2.4	—	2.1	2.1

A family of normalized reverberation spectra, constructed according to Eq. (34.10) for various values of the parameter k, is shown in Fig. 47. The form of the curves bears out the above conclusions regarding the dependence of the shape of the reverberation spectrum on the pulse duration and correla-tion interval of the noise carrier.

Similar results are also obtained for other frequency characteristics on the part of the filters shaping the noise carrier of the signal.

The experimental data on the reverber-ation spectra for the transmission of noisy signals are in good agreement with the results of calculations based on the relations pre-sented above.

Figure 48 shows, as an example, the re-verberation spectra in comparison with the characteristic of the filter shaping the noise carrier.

The measurements were performed for bottom reverberation, with observation of a strong negative refraction of the sound rays in the surface layer to depths of 25-30 m. As the figure shows, given the conditions

$$\Delta F_N\delta \gg 1, \qquad \Delta\omega_N = 2\pi\Delta F_{N'}, \qquad (34.11)$$

the reverberation spectrum is essentially in-dependent of the duration of the transmitted pulses.

Experimental data were also obtained on the width of the reverberation spectrum for various values of the parameter $\Delta F_N\delta$. The experiments were carried out with sur-face reverberation prevailing at the point of reception, using the following procedure.

Using transmitted noise with a constant frequency bandwidth equal to $\Delta F_N = 2.1$ kc, the width ΔF_r of the reverberation spectrum corresponding to different values of the pulse duration between 0.3 and 3 msec was meas-ured. The center frequency of the spectrum was 25 kc.

The values of ΔF_r, like those of ΔF_N, were measured from the peak value of the mean-square spectrum at the 0.7 level.

The results of these measurements are presented in Table 34.1.

It is apparent from the cited data that the width of the reverberation spectrum is practically constant at pulse durations exceeding 5 msec and is determined by the value of ΔF_N.

A similar effect takes place in the transmission of other types of wide-band signals, for example, frequency-modulated signals when the following inequality is satisfied:

$$\Delta F_M \delta \gg 1,$$

like (34.11).

§ 35. General Relations for the Envelope Fluctuation Spectrum

Our next step is to analyze the energy spectra of the reverberation envelope fluctuations.

For our starting relation we use an expression linking the correlation function $B_{rE}(\tau)$ for the envelope fluctuations with the envelope $r_r(\tau)$ of the reverberation correlation coefficient. This expression (see Sec. 24) is approximately written in the form

$$B_{rE}(\tau) \approx \sigma_E^2 r_r^2(\tau), \tag{35.1}$$

where σ_E^2 is the variance of the envelope fluctuations.

Substituting (35.1) into the general relation (32.1), we obtain for the desired spectrum

$$G_{rE}(\omega) \approx 2\sigma_E^2 \int_{-\infty}^{\infty} r_r^2(\tau) \exp(-j\omega\tau)\, d\tau. \tag{35.2}$$

If we then denote by

$$G_{Er}(\omega) = 2 \int_{-\infty}^{\infty} r_r(\tau) \exp(-j\omega\tau)\, d\tau \tag{35.3}$$

the spectrum of the reverberation correlation coefficient envelope, we arrive at the following relation, applying the theorem of convolutions to (35.2) with cognizance of (35.3):

$$G_{rE}(\omega) \approx (\sigma_E^2/4\pi) \int_{-\infty}^{\infty} G_{Er}(v)\, G_{Er}(\omega - v)\, dv. \tag{35.4}$$

Proceeding from (35.4), we calculate the statistical spectrum of the reverberation envelope fluctuations from the known spectrum of the envelope of its correlation coefficient for transmitted signals of any form.

We use this relation for the case when the transmitted signals have a sinusoidal carrier.

The function $r_r(\tau)$ is defined in the form

$$r_r(\tau) = (1/\delta_{ef}) \int_{-\infty}^{\infty} s_0(t)\, s_0(t+\tau)\, dt. \tag{35.5}$$

If now we substitute (35.5) into (35.3), it turns out that

$$G_{Er}(\omega) = (2/\delta_{ef})\, |g_0(\omega)|^2, \tag{35.6}$$

where

$$g_0(\omega) = \int_{-\infty}^{\infty} s_0(t) \exp(-j\omega t)\, dt \qquad (35.7)$$

is the amplitude spectrum of the normalized signal envelope.

Then, taking (35.4) and (35.6) into account, we obtain

$$G_{\text{rE}}(\omega) \approx (\sigma_E^2/\pi\delta_{\text{ef}}^2) \int_{-\infty}^{\infty} |g_0(\nu)|^2 |g_0(\omega-\nu)|^2\, d\nu. \qquad (35.8)$$

Hence, in the given case, <u>the energy spectrum of the fluctuations of the</u> <u>reverberation envelope is proportional to the convolution of the</u> <u>squared amplitude spectra of the normalized envelope of the trans-</u> <u>mitted pulses.</u>

For the transmission of frequency-modulated and noisy signals it is convenient for calculation of the spectra to use the general relation (35.2) or its equivalent

$$G_{\text{rE}}(\omega) \approx 4\sigma_E^2 \int_0^{\infty} r_{\text{r}}^2(\tau) \cos\omega\tau\, d\tau. \qquad (35.9)$$

Here it is necessary first to calculate the reverberation correlation coefficient:

$$R_{\text{r}}(\tau) \approx r_{\text{r}}(\tau) \cos\omega_0\tau,$$

from which the function $r_{\text{r}}(\tau)$ entering into (35.9) is then calculated.

We note that Eq. (35.9) sometimes leads more quickly to the final results in the investigation of pulses with a sinusoidal carrier as well; hence, the use of either of the above analytical relations (35.8) or (35.9) depends only on the extent to which the integration is facilitated.

§ 36. Envelope Fluctuation Spectra for the Transmission of Various Types of Signals

Resting on the results of the preceding section, we now consider some examples of the spectra of reverberation envelope fluctuations for the transmission of pulses with a sinusoidal and wide-band carrier.

If the envelope spectrum $g_0(\omega)$ of a transmitted pulse with sinusoidal carrier has a peak at zero frequency, the following relation holds for the normalized mean-square spectrum $g_{\text{rEn}}(\omega)$ of the reverberation envelope fluctuations, based on (35.8):

$$g_{\text{rEn}}(\omega) \approx \left[\frac{\int_{-\infty}^{\infty} |g_0(\nu)|^2 |g_0(\omega-\nu)|^2\, d\nu}{\int_{-\infty}^{\infty} |g_0(\nu)|^4\, d\nu} \right]^{1/2}. \qquad (36.1)$$

A relation equivalent to (36.1) is obtained from (35.9):

$$g_{\text{rEn}}(\omega) \approx \left[\frac{\int_0^{\infty} r_{\text{r}}^2(\tau) \cos\omega\tau\, d\tau}{\int_0^{\infty} r_{\text{r}}^2(\tau)\, d\tau} \right]^{1/2}, \qquad (36.2)$$

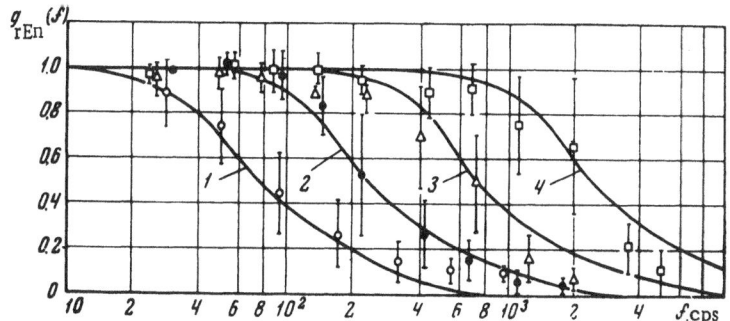

Fig. 49. Calculated (curves) and measured mean-square normalized spectra for fluctuations of the reverberation envelope. 1) $\delta = 10$ msec; 2) $\delta = 3.0$ msec; 3) $\delta = 1.0$ msec; 4) $\delta = 0.3$ msec; $f_0 = 15$ kc.

this being more general than (36.1), insofar as no assumptions have been made with respect to the form of the transmitted signals, other than quasi-harmonicity. In particular, it is possible, by means of (36.2), to find the spectra of the reverberation envelope fluctuations both for pulses with arbitrary form of the sinusoidal carrier envelope, as well as for frequency-modulated and noisy signals.

Let us review some examples.

If the signal has a rectangular envelope and sinusoidal carrier, then, making use of (36.2) and the data of Table 22.1, we obtain

$$g_{rEn}(\omega) \approx (2.45/\omega\delta)\left[1 - \frac{\sin(\omega\delta)}{\omega\delta}\right]^{1/2}. \tag{36.3}$$

Experimental investigations of the spectra for fluctuations of the reverberation envelope obtained under conditions of transmission of rectangular pulses with various durations confirm the dependence (36.3). Figure 49 gives the calculated and measured normalized mean-square spectra for combined surface and bottom reverberation. The procedure used to measure the envelope spectra provided for averaging of the spectral analyzer readings over time (including 10-15 independent readings), as well as over 15-20 reverberation events. The data of Fig. 49 indicate satisfactory agreement between the calculated and experimental values of the spectra.

Another interesting example involves a signal with sinusoidal carrier and a modulated envelope:

$$s_0(t) = \frac{1 + m\cos\Omega_1 t}{1 + m} \exp\left[-(t/t_0)^2\right]. \tag{36.4}$$

The spectrum $g_0(\omega)$ in this case is determined on the basis of (35.7) and (36.4) in the form

$$g_0(\omega) = [\sqrt{\pi}t_0/(1 + m)]\{\exp\left[-(\omega t_0)^2/4\right] +$$
$$+ (m/2)\exp\left[-(\omega - \Omega_1)^2 t_0^2/4\right] + (m/2)\exp\left[-(\omega + \Omega_1)^2 t_0^2/4\right]\}. \tag{36.5}$$

We note that for the normalized spectrum of the transmitted-signal envelope, defined as

$$g_{En}(\omega) = \frac{g_0(\omega)}{g_0(0)},$$

the following relations hold:

$$g_{En}(\omega) = \frac{\exp[-(\omega t_0)^2/4] + (m/2)\exp[-(\omega-\Omega_1)^2 t_0^2/4]}{1 + m\exp[-(\Omega_1 t_0)^2/4]} + \frac{(m/2)\exp[-(\omega+\Omega_1)^2 t_0^2/4]}{1 + m\exp[-(\Omega_1 t_0)^2/4]}. \tag{36.6}$$

The application of (36.1) to (36.5) produces the following results:

$$g_{rEn}(\omega) \approx \left\{ \frac{\exp[-(\omega t_0)^2/4] + (m^2/2)\exp[-(\omega-\Omega_1)^2 t_0^2/4] + (m^2/2)\exp[-(\omega+\Omega_1)^2 t_0^2/4]}{1 + m^2\exp[-(\Omega_1 t_0)^2/4]} \right\}^{1/2}. \tag{36.7}$$

In calculating the relation (36.7), allowance was made for the fact that the values of the products of squared exponentials separated from one another by a frequency interval greatly exceeding the effective width of each turn out to be negligibly small.

A comparison of the normalized spectra (36.6) and (36.7) permit the following conclusions:

The amplitude modulation of the transmitted signal is not completely manifested in the reverberation envelope.

The spectrum of the reverberation envelope fluctuations is broader than the spectrum of the transmitted-signal envelope.

We also note certain features of the spectra of the reverberation envelope in the transmission of wide-band signals.

It was shown in Sec. 25 that with the transmission of signals having a frequency-modulated and noise carrier the correlation characteristics of the reverberation envelope, given the conditions

$$\Delta F_M \delta_{ef} \gg 1, \qquad \Delta F_N \delta_{ef} \gg 1 \tag{36.8}$$

do not depend on the effective pulse duration and are determined by the frequency deviation ΔF_M or the bandwidth of the noise carrier ΔF_N. This implies, in particular, that the reverberation envelope fluctuation spectra, which are uniquely determined by the correlation characteristics, are also independent of the transmitted pulse duration when the conditions (36.8) are fulfilled.

If the law of frequency modulation of the signal or the form of the frequency characteristic shaping the noise carrier is given, the specific form of the reverberation envelope fluctuation spectrum can be found from the relation (36.2).

§ 37. Reverberation Spectra with Motion of the Scatterers and Displacement of the Acoustic Arrays

Any motion of the scatterers in the ocean medium or displacement of the acoustic arrays can alter the spectral characteristics of reverberation significantly in a number of cases. It is interesting in this connection to study certain characteristics of the formation of the spectra for reverberation signals observed under such conditions.

In Sec. 30 the correlation characteristics of reverberation were calculated with allowance for the motion of the scatterers in the ocean. These characteristics make it possible to perform a spectral analysis with fairly general assumptions concerning the distributions of the scatterer velocities.

The statistical spectrum $G_{rs}(\omega)$ in this case may be found on the basis of the general relation (30.15) and (32.3):

$$G_{rs}(\omega) = 4\sigma_s^2 \int\limits_0^\infty r_r(\tau) \sum_{i=1}^N P_i \theta_i(\tau) \cos\left[(\omega_0 + \omega_{si})\tau\right] \cos \omega\tau \, d\tau.$$

From this, considering the quasi-harmonicity of the reverberation, we determine its normalized mean-square spectrum

$$g_{rRn}(\omega) = \left[\frac{\sum\limits_{i=1}^N P_i \int\limits_{-\infty}^\infty G_{Er}(\nu) W_i(\omega_0 - \omega_{si} - \omega - \nu) \, d\nu}{G_{Er}(0)}\right]^{1/2}, \tag{37.1}$$

where $W_i(\Delta\omega)$ is the probability density of the frequency shifts for the i-th set of elementary scattered signals, $G_{r0}(\omega)$ is found from the relations (35.3) or (35.6) and is the Fourier transform of the reverberation correlation envelope. We note that the conditions for normalization of the spectrum in this case are expressed in terms of constant variance regardless of the forms of the distribution $W_i(\Delta\omega)$.

As an example of the formation of a reverberation spectrum we consider the case when the characteristics of the scatterer motion and the transmitted signals are specified in the form (30.18) and (10.8). Making use of these relations, and proceeding from (37.1), we obtain

$$g_{rRn}(\Delta\omega) \approx \left\{P \exp\left[-(\Delta\omega t_0)^2/2\right] + \left[(1-P)/\sqrt{1+(\sigma_\Omega t_0)^2}\right] \exp\left[-(\Delta\omega - \Omega)^2 t_0^2/2\left[1+(\sigma_\Omega t_0)^2\right]\right]\right\}^{1/2},$$

$$\Delta\omega = \omega - \omega_0. \tag{37.2}$$

It follows from an analysis of the relation (37.2) that as the value of the parameter σ_Ω increases, the reverberation spectrum broadens and becomes less symmetrical.

The behavior of the function $g_{rsn}(\omega)$ may be treated analogously for other forms of transmitted signals, distributions $W_i(\Delta\omega)$, and for other values of the parameter P_i.

A second factor influencing the form of the reverberation spectrum is displacement of the acoustic arrays. In this case, assuming that the scattering of sound occurs at inhomogeneities concentrated in a thin layer, we obtain the following relation for the energy spectrum on the basis of (31.6) and (32.3):

$$G_{rA}(\omega) = 2\langle n_1\rangle\langle a^2\rangle \delta_{ef} \int\limits_0^\infty \int\limits_0^{2\pi} r_r(\tau) \varphi_T^2(\alpha - \alpha_T) \varphi_R^2(\alpha - \alpha_R) W(\alpha) \times$$
$$\times \cos\left[\omega_0(1 + 2v_A \cos\alpha/c)\tau\right] \cos\omega\tau \, d\alpha \, d\tau.$$

It is also possible to go over to the normalized mean-square spectrum g_{rAn}. Bearing in mind the quasi-harmonicity of the reverberation and the relation (35.3), we obtain

$$g_{rAn}(\omega) \approx \left\{\frac{\int\limits_0^{2\pi} \varphi_T^2(\alpha - \alpha_T)\varphi_R^2(\alpha - \alpha_R) W(\alpha) G_{Er}[\omega_0(1 + 2v_A\cos\alpha/c) - \omega] \, d\alpha}{G_{Er}(0)\int\limits_0^{2\pi} \varphi_T^2(\alpha - \alpha_T)\varphi_R^2(\alpha - \alpha_R) W(\alpha) \, d\alpha}\right\}^{1/2}, \tag{37.3}$$

the normalization of the spectrum being performed here as in the derivation of Eq. (37.1).

For our subsequent calculations of the reverberation spectra, we need to specify the form of the distribution $W(\alpha)$ of the scatterers in space and the directivity characteristic of the transmitter $\varphi_T(\alpha)$ and receiver $\varphi_R(\alpha)$, as well as the function $G_{Er}(\omega)$, which is determined by the type of signals transmitted.

We now present another useful relation for the reverberation spectrum for the case of nondirectional transmission and directional reception.

If, as before, the scatterer distribution is assumed to be spatially uniform, we obtain from (37.3)

$$g_{rAn}(\omega) \approx \left\{ \frac{\int_0^{2\pi} \varphi_R^2(\alpha - \alpha_R)\, G_{Er}[\omega_0(1 + 2v_A \cos \alpha/c) - \omega]\, d\alpha}{G_{Er}(0) \int_0^{2\pi} \varphi_R^2(\alpha - \alpha_R)\, d\alpha} \right\}^{1/2} . \tag{37.4}$$

In conclusion, we find an expression for the reverberation spectrum with allowance for the combined motion of the scatterers in the ocean and displacement of the acoustic arrays. For this purpose we make use of the generalization of the two-dimensional theorem of superposition of stochastic perturbations (see Sec. 7) and perform statistical averaging over $\Delta\omega$ and α.

If the distribution of the frequency shifts for the scatterers is described by an even function $W(\Delta\omega)$, while their distribution in space is given by the probability density $W(\alpha)$, the autocorrelation function of the reverberation then turns out on the basis of (7.3), to be equal to

$$B_{rAs}(\tau) \approx (\langle n_1 \rangle \langle a^2 \rangle / 2) \int_{-\infty}^{\infty} W(\Delta\omega) \cos(\Delta\omega\tau)\, d(\Delta\omega) \times$$

$$\times \int_0^{2\pi} \varphi_T^2(\alpha - \alpha_T)\, \varphi_R^2(\alpha - \alpha_R)\, W(\alpha) \cos[\omega_0(1 + 2v_A \cos \alpha/c)\, \tau]\, d\alpha \times$$

$$\times \int_{-\infty}^{\infty} s_0(t)\, s_0(t + \tau) \cos[\Phi(t + \tau) - \Phi(t)]\, dt. \tag{37.5}$$

We next account for the fact that the reverberation variance σ_V^2 in the given instance is equal to

$$\sigma_V^2 = (\langle n_1 \rangle \langle a^2 \rangle \delta_{ef}/2) \int_{-\infty}^{\infty} \varphi_T^2(\alpha - \alpha_T)\, \varphi_R^2(\alpha - \alpha_R)\, W(\alpha)\, d\alpha. \tag{37.6}$$

Then, with the relations (30.5) and (31.7) in mind, we write the correlation function (37.5) in the form

$$B_{rAs}(\tau) = \sigma_V^2 \theta(\tau)\, R_{rA}(\tau). \tag{37.7}$$

Consequently, the reverberation correlation function is expressed in terms of the product of the characteristic function $\Theta(\tau)$ of the distribution $W(\Delta\omega)$ of the frequency shifts elicited by scatterer motion, and the correlation coefficient $R_{rA}(\tau)$, which depends on the form of the transmitted signals, the profile of the directivity patterns of the acoustic arrays, and their velocity of motion.

Making use of (32.3) and (37.7), we obtain the following expression for the energy spectrum $G_{rAs}(\omega)$:

$$G_{rAs}(\omega) \approx (\sigma_V^2/4\pi) \int_{-\infty}^{\infty} W(v)\, G_{rA}(\omega - v)\, dv. \tag{37.8}$$

It follows from (37.8) that the reverberation spectrum, taking into account the combined influence of motion of the scatterers and acoustic arrays, is the convolution of the probability density $W(\Delta\omega)$ and the spectrum

$$G_{rA}(\omega) = 4 \int_0^\infty R_{rA}(\tau) \cos \omega\tau \, d\tau. \tag{37.9}$$

CHAPTER VI

ADDITIONAL PROBLEMS IN THE ANALYSIS OF REVERBERATION SIGNALS

§38. Joint Space – Time Correlation

In the solution of problems associated with the action of reverberation signals on the acoustic arrays and other forms of instrumentation, it often becomes necessary to regard the space–time statistical relations in the scattered fields.

Such relations may be described, for example, by means of the space–time correlation functions, which we investigate under the following assumptions.

We will assume that the transmitting array has a directivity characteristic $\varphi_T(\alpha, \theta)$, and that reception is realized by two nondirectional point hydrophones situated at a distance r apart (Fig. 50). We assume further that the reverberation is caused by scattering by inhomogeneities concentrated in a layer of arbitrary thickness h, with the distribution of the scatterer density in the layer thickness described by the function $W_z(z)$.

The scatterers are able to move in the general case, and we will describe the random displacements of the carrier frequency of the elementary scattered signals by a certain distribution law $W(\Delta\omega)$.

To calculate the joint space–time correlation function, we use the generalization of the two-dimensional theorem of superposition of stochastic perturbations (see Sec. 7). Applying the relation (7.3), we perform statistical averaging of the product $V_1(t)V_2(t)$ over the set of stochastic parameters, the role of which, in this case, is taken by the angles α, θ, and the frequency shift $\Delta\omega$. As a result of this averaging, we obtain the sought-after correlation function in the case when the ocean medium has uniform scattering properties in the horizontal plane

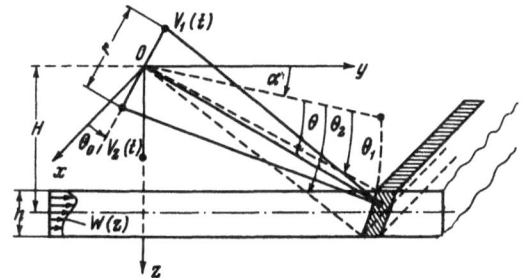

Fig. 50. Coordinate system used in calculating the space–time correlation of reverberation.

$$B_{12}(\tau, r) = (\langle n_1 \rangle \langle a^2 \rangle / 2\pi) \int\limits_0^{2\pi} \int\limits_{\theta_1}^{\theta_2} \varphi_T^2(\alpha, \theta) \, W_z(\sin\theta) \times$$

$$\times \int\limits_{-\infty}^{\infty} W(\Delta\omega) \int\limits_{-\infty}^{\infty} s_0(t) \, s_0(t+\tau) \times$$

$$\times \cos\{(\omega_0 + \Delta\omega)[t + \Delta t(\alpha, \theta)/2] + \Phi(t)\} \times$$

$$\times \cos\{(\omega_0 + \Delta\omega)[t + \tau - \Delta t(\alpha, \theta)/2] +$$

$$+ \Phi(t+\tau)\} \, dt \, d(\Delta\omega) \cos\theta \, d\theta \, d\alpha. \qquad (38.1)$$

For simplicity in formulating the expression (38.1), the distance from the transmitter to the scattering volume was assumed equal to unity, and, consistent with the adopted coordinate system, the following relation was borne in mind:

$$W_z (\sin \theta) \cos \theta \, d\theta = W_z (z) dz,$$
$$\theta_1 = \arcsin (H - h/2),$$
$$\theta_2 = \arcsin (H + h/2),$$
(38.2)
$$\Delta t(\alpha, \theta) \approx (r/c) (\cos \theta_0 \cos \theta \sin \alpha + \sin \theta_0 \sin \theta).$$

With regard also for the quasi-harmonicity of the transmitted signals and the assumed evenness of the function $W(\Delta\omega)$, we go from (38.1) to the relation

$$B_{12} (\tau, r) \approx (\langle n_1 \rangle \langle a^2 \rangle / 4\pi) R_{\mathrm{r}}(\tau) \int_0^{2\pi} \int_{\theta_1}^{\theta_2} \varphi_{\mathrm{T}}^2 (\alpha, \theta) W_z (\sin \theta) \times$$
$$\times \dot\Theta[\tau + \Delta t (\alpha, \theta)] \cos [\omega_0 \Delta t (\alpha, \theta)] \cos \theta \, d\theta \, d\alpha,$$
(38.3)

where

$$R_{\mathrm{r}} (\tau) \approx r_{\mathrm{r}} (\tau) \cos \omega_0 \tau$$
(38.4)

is the reverberation autocorrelation coefficient, calculated in the absence of scatterer motion, and

$$\Theta (\eta) = 2 \int_0^\infty W (\Delta\omega) \cos (\eta \Delta\omega) \, d (\Delta\omega)$$
(38.5)

is the characteristic function of the distribution $W(\Delta\omega)$.

The space–time correlation coefficient, defined as

$$R_{12} (\tau, r) = \frac{B_{12} (\tau, r)}{B_{12} (0, 0)} ,$$

may be written in the following form on the basis of (38.3):

$$R_{12}(\tau, r) \approx \frac{R_{\mathrm{r}}(\tau) \int_0^{2\pi} \int_{\theta_1}^{\theta_2} \varphi_{\mathrm{T}}^2 (\alpha, \theta) W_z (\sin\theta) \Theta[\tau + \Delta t(\alpha, \theta)] \cos [\omega_0 \Delta t(\alpha, \theta)] \cos \theta \, d\theta \, d\alpha}{\int_0^{2\pi} \int_{\theta_1}^{\theta_2} \varphi_{\mathrm{T}}^2 (\alpha, \theta) W_z (\sin\theta) \cos \theta \, d\theta \, d\alpha} .$$
(38.6)

The expressions derived above for the space–time statistical characteristics of reverberation make it possible to study the properties of the scattered sound fields for various spatial density distributions of the scatterers and various characteristics of their motion. We notice that the expressions given in Secs. 27 and 30 for the time and spatial correlation characteristics of reverberation follow, in particular, from (38.6).

As one example of the application of the general relation (38.6), we examine the spatial correlation for sound scattering by inhomogeneities concentrated in a thin layer.

We assume here that

$$\tau = 0, \quad \varphi_{\mathrm{T}}(\alpha, \theta) = 1,$$
$$W_z (\sin \theta) = 1/(\theta_2 - \theta_1), \quad \theta_2 - \theta_1 \ll 1,$$
(38.7)
$$\theta_0 = 0, \quad H = 0.$$

Then, taking (38.6) and (38.7) into account, we obtain for the spatial correlation coefficient

$$B_{12}(0, r) \approx (1/2\pi) \int_0^{2\pi} \Theta\,[(r/c)\sin\alpha]\cos(kr\sin\alpha)\,d\alpha. \qquad (38.8)$$

With scatterer motion absent,

$$\theta\,[(r/c)\sin\alpha] = 1,$$

and we arrive at the relation (27.17).

If we include the motion of the scatterers and assume, for example, that the frequency shift is described by a normal distribution, i.e.,

then

$$W(\Delta\omega) = (1/\sqrt{2\pi}\sigma_\Omega)\exp[-(\Delta\omega^2/2\sigma_\Omega^2)], \quad \sigma_\Omega \approx 2\sigma_s^2\omega_0/c,$$

$$\Theta\,[(r/c)\sin\alpha] = \exp[-2(\sigma_s/c)^2(kr)^2\sin^2\alpha],$$

where σ_s is the rms velocity of the moving scatterers. The correlation coefficient (38.8) in this case may be written in the form

$$R_{12}(0, r) \approx \int_0^{2\pi} \exp[-2(\sigma_s/c)^2(kr)^2\sin^2\alpha]\cos(kr\sin\alpha)\,d\alpha. \qquad (38.9)$$

If the following inequality is valid:

$$kr\,(\sigma_s/c) \ll 1,$$

which is satisfied in practice to values of σ_s equal to several meters per second, and kr is on the order of tens, we are in a position to write approximately

$$\exp[-2(\sigma_s/c)^2(kz)^2\sin^2\alpha] \approx 1 - 2(\sigma_s/c)^2(kr)^2\sin^2\alpha. \qquad (38.10)$$

Then, taking (38.10) into account, the expression for the correlation coefficient (38.9) reduces to the following form after integration:

$$R_{12}(0, r) \approx J_0(kr)\,[1 - (\sigma_s/c)^2(kr)^2] + \pi(\sigma_s/c)^2(kr)^2 J_2(kr). \qquad (38.11)$$

Hence, it follows that the spatial correlation of reverberation is reduced when the scatterers are in motion, the degree of this reduction depending on the parameter σ_s/c characterizing the relative rms velocity of the moving scatterers.

Analogously, making use of the above relations, one can estimate the space–time correlation of reverberation for other distributions $W(\Delta\omega)$ and $W_z(z)$ and other transmitter directivity characteristics $\varphi_T(\alpha, \theta)$, as well.

§ 39. Ambiguity Functions and Diagrams

The modulation correlation function $C(\tau, \Omega)$ was determined for transmitted signals s(t) in Sec. 12, where also some of its properties were mentioned. It was recognized, in particular, that a convenient measure of the joint frequency–time correlation is found in the ambiguity diagrams, i.e., figures generated by the intersection of the surface $|C(\tau, \Omega)|$ with the plane

$$|C(\tau, \Omega)| = C_0, \qquad C_0 = \text{const.} \tag{39.1}$$

This approach to the analysis of the joint frequency—time statistical relations proves useful in studying the properties of reverberation signals.

Let us compare the modulation correlation function of the transmitted signals (12.1) with the reverberation frequency—time correlation coefficient $r_r(\tau, \Omega)$ determined in (29.4). It is evident that the following equation holds for them:

$$|C(\tau, \Omega)| = r_r(\tau, \Omega). \tag{39.2}$$

It may be inferred from this that the s p a c e – t i m e   c o r r e l a t i o n   c o e f f i c i e n t   o f r e v e r b e r a t i o n   c o i n c i d e s   w i t h   t h e   m o d u l u s   o f   t h e   j o i n t   m o d u l a t i o n   c o r - r e l a t i o n   f u n c t i o n   o f   t h e   t r a n s m i t t e d   s i g n a l s .   This permits the concept of reverberation ambiguity diagrams and functions to be introduced and the ambiguity diagrams to be found from the known characteristics of the transmitted signals.

Consequently, for calculation of the reverberation ambiguity diagrams we use the following relation:

$$|C(\tau, \Omega)| = (1/\delta_{ef}) \left| \int_{-\infty}^{\infty} s_0(t) s_0(t + \tau) \ \exp\{j[\Omega t + \Phi(t+\tau) - \Phi(t)]\} dt \right|, \tag{39.3}$$

which is valid for the transmission of determinate signals with an arbitrary amplitude and frequency modulation configuration.

As an example we consider the case of signal transmission with a sinusoidal carrier and bell-shaped (Gaussian) envelope:

$$s_0(t) = \exp[-(t/t_0)^2], \quad \delta_{ef} = \sqrt{\pi/2}t_0. \tag{39.4}$$

The modulus of the ambiguity function, calculated according to Eq. (39.3), turns out, after substitution therein of (39.4) and integration, to be equal to

$$|c(\tau, \Omega)| = \exp[-(\pi\tau^2/4\delta_{ef}^2) - (\Omega\delta_{ef})^2/4\pi]. \tag{39.5}$$

Next, following the procedure outlined in Sec. 12, we determine the ambiguity diagram corresponding to the investigated signal. The intersection of the surface $|C(\tau, \Omega)|^2$ with a plane parallel to the coordinate axes (Ω, τ) is chosen at that level C_0^2 for which the volume of a cylinder whose base is equal to the area of intersection is equal to the volume of the ambiguity body.

The ambiguity diagram so determined is described by the equation

$$(\pi\tau^2/2\delta_{ef}^2) + (\Omega\delta_{ef})^2/2\pi = 1, \tag{39.6}$$

where

$$C_0 = \exp(-\,^1/_2). \tag{39.7}$$

Equation (39.6) is the equation of an ellipse with semiaxes

$$\left.\begin{array}{l} a = \sqrt{2/\pi}\delta_{ef}, \\ b = \sqrt{2\pi}/\delta_{ef}, \end{array}\right\} \tag{39.8}$$

which characterize the correlations with respect to time and frequency, respectively.

Let us investigate one further example, when the transmitted signal has a bell-shaped envelope (39.4) and is frequency-modulated by a linear time function, so that

$$s_{01}(t) = \exp\left[(j\pi\Delta F_{\text{M}}t^2/2\delta_{\text{ef}}^2) - (\pi t^2/2\delta_{\text{ef}}^2)\right], \quad \Delta\omega_{\text{M}} = 2\pi\Delta F_{\text{M}}. \tag{39.9}$$

The modulus of the ambiguity function in this case is described by the expression

$$|C(\tau, \Omega)| = \exp\left\{-\left[\pi\Delta F_{\text{M}}^2 + (\pi/4\,\delta_{\text{ef}}^2)\right]\tau^2 - \Delta F_{\text{M}}\delta_{\text{ef}}\Omega\tau - (\Omega\delta_{\text{ef}})^2/4\pi\right\}, \tag{39.10}$$

and the equation for the ambiguity diagram has the form

$$\left[2\pi\Delta F_{\text{M}}^2 + (\pi/2\,\delta_{\text{ef}}^2)\right]\tau^2 + 2\Delta F_{\text{M}}\delta_{\text{ef}}\Omega\tau + (\Omega\delta_{\text{ef}})^2/2\pi = 1. \tag{39.11}$$

We note that the parameter C_0 in this case coincides with the value indicated in (39.7).

Equation (39.11), like (39.6), is the equation of an ellipse. The only difference is that the ellipse described by (39.11) is deformed along its axis and is rotated relative to the ellipse described by (39.6) through an angle α equal to

$$\alpha = (^1/_2)\,\text{arctg}\,\left[4\pi\Delta F_{\text{M}}\delta_{\text{ef}}^2/(4\pi\Delta F_{\text{M}}^2\delta_{\text{ef}}^2 + \pi^2\delta_{\text{ef}}^4)\right]. \tag{39.12}$$

It is easily seen that for $\Delta F_{\text{M}} = 0$ the angle $\alpha = 0$, and Eq. (39.11) goes over to (39.6).

The property of the ambiguity diagram that its form is independent of the magnitude of the frequency deviation is clearly inherent in signals with other types of envelope as well. As another example we consider a signal with an exponential envelope

$$s_0(t) = \exp(-t/t_0), \; t \geqslant 0, \; \delta_{\text{ef}} = t_0/2 \tag{39.13}$$

and sinusoidal carrier.

The modulus of the ambiguity function in this case is determined by the relation

$$|C(\tau, \Omega)| = \exp(-|\tau|/2\delta_{\text{ef}})/[1 + (\Omega\delta_{\text{ef}})^2]^{1/2}, \tag{39.14}$$

where the parameter C_0 is equal to

$$C_0 = {}^1/_2\sqrt{3}. \tag{39.15}$$

The equation for the ambiguity diagram is written in the form

$$(\Omega\delta_{\text{ef}})^2 = 12\exp(-|\tau|/\delta_{\text{ef}}) - 1. \tag{39.16}$$

We examine still another example of transmitted signals, when the latter represent noisy pulses. In this case it is meaningful to talk about the average modulation correlation function $<C(\tau, \Omega)>$ (see Sec. 12).

Let the determinate envelope of the signal $s_0(t)$ be described by the squared exponential (39.4), the envelope of the correlation coefficient $r_x(\tau)$ of the noise carrier by the expression

$$r_x(\tau) = \exp(-\pi\tau^2/4\tau_{\text{N}}^2), \tag{39.17}$$

where τ_{N} is the correlation interval of the noise carrier. In the above form, $r_x(\tau)$ corresponds to the frequency characteristic of a filter producing noise in the form of a squared exponential (see Table 23.1, row 2).

On the basis of the relations (12.18) and (39.3)-(39.5), we obtain for $<C(\tau, \Omega)>$

$$\langle C(\tau, \Omega) \rangle = \exp\{-(\pi/4)[(\delta_{ef}/\tau_N)^2 + 1](\tau/\delta_{ef})^2 - (\Omega\delta_{ef})^2/4\pi\}. \tag{39.18}$$

The ambiguity diagram determined from (39.18) is described by the equation

$$(\pi/2)[(\delta_{ef}/\tau_N)^2 + 1](\tau/\delta_{ef})^2 + (\Omega\delta_{ef})^2/2\pi = 1. \tag{39.19}$$

It is apparent from the above equation, in particular, that if the noise carrier is a narrow-band type and

$$\tau_N \gg \delta_{ef},$$

then (39.19) transforms to (39.6). This corresponds to the transmission of a signal with a sinusoidal carrier. On the other hand, for

$$\tau_N \ll \delta_{ef}$$

we have, from (39.19),

$$(\pi/2)(\tau/\tau_N)^2 + (\Omega\delta_{ef})^2/2\pi \approx 1. \tag{39.20}$$

It is apparent from this that the time correlation depends only on the properties of the noise carrier, the frequency correlation only on the effective pulse duration.

This is an interesting property of noise signals, whereby the ambiguity diagram is deformed only along the τ axis as the frequency band of the noise carrier is expanded.

The ambiguity diagrams for other types of signals are found analogously.

It is important to realize that the method of analyzing the properties of signals and reverberation by means of their ambiguity diagrams is a fairly general technique for studying the statistical characteristics of these processes. Such a method makes it possible, in particular, to investigate the complex problems associated with the resolving power of various underwater acoustic devices.

The investigation of the properties of various types of signals in the aspect considered here is described in a number of papers (see [3, 53]). In particular [53] is devoted to the analysis of sonar systems from the point of view of enhancing their resolving power and resistance to noise.

It is emphasized in this paper that one of the most important problems involved in signal detection in the broad sense is to find the transmitted signal form for which it will be possible to determine most accurately both the distance to the object in question, as well as the velocity of its motion. This means, for example, that in the correlation method one must have a transmitted signal form for which the region of ambiguity for the received signal will overlap minimally with the ambiguity region of reverberation noise.

The investigation of ambiguity diagrams for signals with a sinusoidal and frequency-modulated carrier, as well as for noisy signals motivated the following conclusions on the part of the authors of the given paper:

The optimum receiver of sonar signals against noise and interference having a normal distribution should form at its output the cross-correlation function of the received signal, reproducing the transmitted signal with a time and frequency shift.

Short, frequency-modulated, and noisy signals create a high resolving power with respect to the time (i.e., to the distance).

Long signals with a sinusoidal carrier create a high resolving power with respect to the frequency.

The solution of combination sonar detection problems requiring the measurement of two or more parameters of the received signals can be achieved with the transmission of several types of signals, each of which permits the realization of a high resolving power with respect to the appropriate combination of measured parameters.

We point out, in conclusion, that the reverberation ambiguity functions and diagrams, in general, depend on the characteristics of the motion of the scatterers and the displacement of the acoustic arrays. Adjustment for the influence of these factors may be made on the basis of the generalization of the theorem of superposition of stochastic perturbations, provided the joint frequency–time correlation function of the reverberation is calculated with statistical averaging of the elementary scattered signals over all possible stochastic parameters; of course, the probability distributions of these parameters must be known in this case.

§ 40.   Phase-Difference Fluctuations of Reverberation Signals

The information obtained above with regard to the statistical properties of reverberation enables us to formulate and solve the following problem: to find the phase-difference distribution of two stationary quasi-harmonic reverberation processes $V_1(t)$ and $V_2(t)$ under the condition that their cross-correlation coefficient is specified. This problem can be solved if the two-dimensional distribution is known for the given processes.

The model of reverberation as a stochastic process and the analysis of the distributions of its instantaneous values (see Secs. 4, 13) foster the assumption that the two-dimensional probability density $W(V_1, V_2)$ of the two correlated processes $V_1(t)$ and $V_2(t)$ obeys a normal law:

$$W(V_1, V_2) = \frac{1}{2\pi\sigma_{V_1}\sigma_{V_2}(1-R^2)^{1/2}} \exp\left[-\frac{V_1^2 + V_2^2 - 2RV_1V_2}{2\sigma_{V_1}\sigma_{V_2}(1-R^2)}\right] \tag{40.1}$$

or one similar to it. In Eq. (40.1) the symbols σ_{V_1} and σ_{V_2} denote the rms values of the indicated processes, R is their cross-correlation coefficient, which has the sense, for example, of the coefficient of spatial, frequency, time, or joint space–time correlation.

We now proceed with the solution of the stated problem.

Under the postulated assumptions, the processes $V_1(t)$ and $V_2(t)$ may be written in the form

$$\left.\begin{aligned} V_1(t) &= V_{c1}(t)\cos\omega_0 t - V_{s1}(t)\sin\omega_0 t, \\ V_2(t) &= V_{c2}(t)\cos\omega_0 t - V_{s2}(t)\sin\omega_0 t, \end{aligned}\right\} \tag{40.2}$$

where $V_{c1}(t)$, $V_{s1}(t)$, $V_{c2}(t)$, and $V_{s2}(t)$ are the quadrature components of the reverberation signals (see Sec. 16).

Here, the instantaneous phases $\psi_1(t)$ and $\psi_2(t)$ of the investigated processes are defined as follows:

$$\left.\begin{aligned} \psi_1(t) &= \operatorname{arctg}[V_{s1}(t)/V_{c1}(t)], \\ \psi_2(t) &= \operatorname{arctg}[V_{s2}(t)/V_{c2}(t)]. \end{aligned}\right\} \tag{40.3}$$

For the calculation of the phase-difference probability density it is necessary to write the joint distribution $W_V(V_{c1}, V_{s1}, V_{c2}, V_{s2})$ of the quadrature components. This distribution is normal, by virtue of (40.1), so that

$$W_V (V_{c1}, V_{s1}, V_{c2}, V_{s2}) = \frac{1}{(2\pi)^2 \sigma_{V_1}^2 \sigma_{V_2}^2 |K|^{1/2}} \exp - \left(\frac{\sum_{i=1}^{4} \sum_{j=1}^{4} V_{ci} V_{sj} |K_{ij}|}{2\sigma_{V_1} \sigma_{V_2} |K|} \right),$$

(40.4)

where $|K|$ is the determinant of the correlation matrix of the quadrature components, $|K_{ij}|$ is its signed minor.

The specific form of the distribution (40.4) is obtained when the elements of the correlation matrix are known:

$$K = \begin{Vmatrix} R_{11} & R_{12} & R_{13} & R_{14} \\ R_{21} & R_{22} & R_{23} & R_{24} \\ R_{31} & R_{32} & R_{33} & R_{34} \\ R_{41} & R_{42} & R_{43} & R_{44} \end{Vmatrix},$$

(40.5)

which may be determined as follows. If we use the relation (40.2) and find the quantities

$$\langle V_1(t) V_2(t) \rangle / \sigma_{V_1} \sigma_{V_2}; \quad \langle V_1^2(t) \rangle / \sigma_{V_1}^2; \quad \langle V_2^2(t) \rangle / \sigma_{V_2}^2,$$

assuming also stationary coupling between the processes $V_1(t)$ and $V_2(t)$, we obtain, as a result, the following set of equations:

$$\left. \begin{aligned} &\langle V_{c1}(t) V_{c2}(t) \rangle = \langle V_{s1}(t) V_{s2}(t) \rangle, \\ &\langle V_{c1}(t) V_{c2}(t) \rangle + \langle V_{s1}(t) V_{s2}(t) \rangle = 2R\sigma_{V_1}\sigma_{V_2}, \\ &\langle V_{s1}(t) V_{c2}(t) \rangle = - \langle V_{c1}(t) V_{s2}(t) \rangle, \\ &\langle V_{c1}(t) V_{s1}(t) \rangle = 0, \\ &\langle V_{c2}(t) V_{s2}(t) \rangle = 0, \\ &\langle V_{c1}^2(t) \rangle = \langle V_{s1}^2(t) \rangle, \\ &\langle V_{c2}^2(t) \rangle = \langle V_{s2}^2(t) \rangle, \\ &\langle V_{c1}^2(t) \rangle + \langle V_{s1}^2(t) \rangle = 2\sigma_{V_1}^2, \\ &\langle V_{c2}^2(t) \rangle + \langle V_{s2}^2(t) \rangle = 2\sigma_{V_2}^2. \end{aligned} \right\}$$

(40.6)

Solving the system of equations (40.6), it can be shown that the correlation matrix (40.5) is reducible to the form

$$K = \begin{Vmatrix} 1 & 0 & R & S \\ 0 & 1 & -S & R \\ R & -S & 1 & 0 \\ S & R & 0 & 1 \end{Vmatrix},$$

(40.7)

where

$$R = \langle V_1(t) V_2(t) \rangle / \sigma_{V_1} \sigma_{V_2}$$

(40.8)

is the cross-correlation coefficient of the given reverberation processes, and the parameter S is the cross-correlation coefficient of the distinct quadrature components:

$$S = \langle V_{c1}(t) V_{s2}(t) \rangle / \sigma_{V_1} \sigma_{V_2} = - \langle V_{c2}(t) V_{s1}(t) \rangle / \sigma_{V_1} \sigma_{V_2}.$$

(40.9)

The determinant $|K|$ and the signed minors $|K_{ij}|$ of the correlation matrix (40.7) turn out to be equal to

$$\left.\begin{aligned}
&|\boldsymbol{K}| = (1 - R^2 - S^2)^2, \\
&|\boldsymbol{K}_{11}| = |\boldsymbol{K}_{22}| = |\boldsymbol{K}_{33}| = |\boldsymbol{K}_{44}| = |\boldsymbol{K}|^{1/2}, \\
&|\boldsymbol{K}_{12}| = |\boldsymbol{K}_{21}| = |\boldsymbol{K}_{34}| = |\boldsymbol{K}_{43}| = 0, \\
&|\boldsymbol{K}_{13}| = |\boldsymbol{K}_{31}| = |\boldsymbol{K}_{24}| = |\boldsymbol{K}_{42}| = -R\,|\boldsymbol{K}|^{1/2}, \\
&|\boldsymbol{K}_{14}| = |\boldsymbol{K}_{41}| = -|\boldsymbol{K}_{23}| = -|\boldsymbol{K}_{32}| = -S\,|\boldsymbol{K}|^{1/2}.
\end{aligned}\right\} \tag{40.10}$$

Substituting (40.10) into the relation (40.4), we obtain

$$W_V(V_{c1}, V_{s1}, V_{c2}, V_{s2}) = \frac{1}{2\pi\sigma_{V_1}^2\sigma_{V_2}^2(1 - R^2 - S^2)} \times$$

$$\times \exp\left[-\frac{V_{c1}^2 + V_{s1}^2 + V_{c2}^2 + V_{s2}^2 - 2R(V_{c1}V_{c2} + V_{s1}V_{s2}) - 2S(V_{c1}V_{s2} - V_{s1}V_{c2})}{2\sigma_{V_1}\sigma_{V_2}(1 - R^2 - S^2)}\right]. \tag{40.11}$$

We next find the joint distribution $W(E_1, E_2, \psi_1, \psi_2)$ of the envelopes $E_1(t)$, $E_2(t)$, and the instantaneous phases $\psi_1(t)$, $\psi_2(t)$ of the investigated processes, applying the general rule

$$\left.\begin{aligned}
&W(E_1, E_2, \psi_1, \psi_2) = E_1 E_2 W_V(E_1\cos\psi_1, E_1\sin\psi_1, E_2\cos\psi_2, E_2\sin\psi_2) \\
&E_1, E_2 \geqslant 0, \quad 0 \leqslant \psi_1, \quad \psi_2 < 2\pi.
\end{aligned}\right\} \tag{40.12}$$

After substitution of (40.11) into (40.12) we arrive at the expression

$$W(E_1, E_2, \psi_1, \psi_2) = \frac{E_1 E_2}{(2\pi)^2\sigma_{V_1}^2\sigma_{V_2}^2(1 - R^2 - S^2)}\ \exp\left[-\frac{E_1^2 + E_2^2 - 2(R^2 + S^2)E_1 E_2\cos(\psi_2 - \psi_1 - \varphi)}{2\sigma_{V_1}\sigma_{V_2}(1 - R^2 - S^2)}\right], \tag{40.13}$$

where φ denotes the following:

$$\varphi = \tan^{-1}(S/R). \tag{40.14}$$

The joint probability density $W(E_1, E_2, \Delta\psi)$ of the envelopes and phase difference is obtained from (40.13):

$$\Delta\psi(t) = \psi_2(t) - \psi_1(t)$$

by substitution of the variables in Eq. (40.13) according to the formulas

$$\Delta\psi = \psi_2 - \psi_1, \quad \psi = \psi_1$$

and integration over ψ. We have, as a result,

$$W(E_1, E_2, \Delta\psi) = \frac{E_1 E_2}{2\pi\sigma_{V_1}^2\sigma_{V_2}^2(1 - R^2 - S^2)}\ \exp\left[-\frac{E_1^2 + E_2^2 - 2(R^2 + S^2)^{1/2}E_1 E_2\cos(\Delta\psi - \varphi)}{2\sigma_{V_1}\sigma_{V_2}(1 - R^2 - S^2)}\right]. \tag{40.15}$$

The expression (40.15) can be used to find the joint distribution $W(E_1, E_2)$ of the envelopes and the one-dimensional distribution of the phase difference $W(\Delta\psi)$. In order to obtain the phase-difference probability density of the reverberation processes, which is what we are after, it is necessary to integrate $W(E_1, E_2, \Delta\psi)$ over all possible values of the variables E_1 and E_2, i.e.,

$$W(\Delta\psi) = \int_0^\infty\int_0^\infty W(E_1, E_2, \Delta\psi)\,dE_1\,dE_2. \tag{40.16}$$

Substitution of (40.15) into (40.16) produces the following integral representation for $W(\Delta\psi)$:

$$W(\Delta\psi) = \frac{1}{2\pi\sigma_{V_1}^2\sigma_{V_2}^2(1 - R^2 - S^2)}\ \int_0^\infty\int_0^\infty E_1 E_2\exp\left[-\frac{E_1^2 + E_2^2 - 2(R^2 + S^2)E_1 E_2\cos(\Delta\psi - \varphi)}{2\sigma_{V_1}\sigma_{V_2}(1 - R^2 - S^2)}\right]dE_1\,dE_2.$$

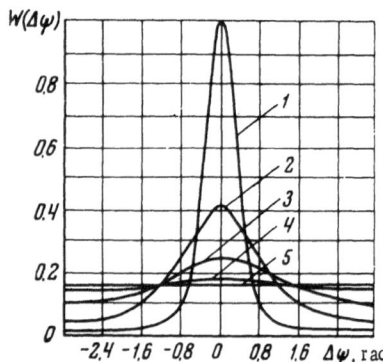

Fig. 51. Probability density of the phase difference of reverberation signals for various values of the cross-correlation coefficient. 1) R = 0.9; 2) 0.6; 3) 0.3; 4) 0.1; 5) 0.

If in the integral so obtained we replace the variables as follows:

$$E_1 = [\sigma_{V_1}\sigma_{V_2}(1 - R^2 - S^2)\,x]^{1/2}\exp(y/2),$$

$$E_2 = [\sigma_{V_1}\sigma_{V_2}(1 - R^2 - S^2)\,x]^{1/2}\exp(-y/2)$$

and allow for the fact that

$$\mathcal{K}_0(x) = \int\limits_0^\infty \exp(-x\,\text{ch}\,y)\,dy,$$

we can reduce it to the form

$$W(\Delta\psi) = \frac{1 - R^2 - S^2}{2\pi}\int\limits_0^\infty x\exp\{(R^2 + S^2)^{1/2}\times$$

$$\times \cos[x(\Delta\psi - \varphi)]\}\,\mathcal{K}_0(x)\,dx,$$

where $\mathcal{K}_{0\,0}(x)$ is a zero-order cylindrical harmonic of an imaginary argument.

Then, making use of the integral [9, 12]

$$\int\limits_0^\infty x\exp(\beta x)\,\mathcal{K}_0(x)\,dx = (1 - \beta^2)^{-3/2}\,[\beta\,\text{arc sin }\beta + (\pi\beta/2) + (1 - \beta^2)^{1/2}];$$

$$|\beta| \leqslant 1,$$

we arrive at the final expression for the probability density of the phase difference:

$$\left.\begin{aligned}W(\Delta\psi) &= \frac{1 - R^2 - S^2}{2\pi(1 - \beta^2)^{3/2}}\,[\beta\arcsin\beta + (\pi\beta/2) + (1 - \beta^2)^{1/2}],\\[4pt]|\Delta\psi - \varphi| &\leqslant \pi, \quad \beta = (R^2 + S^2)^{1/2}\cos(\Delta\psi - \varphi).\end{aligned}\right\} \tag{40.17}$$

The distribution (40.17), which defines the fluctuations of the phase difference of reverberation signals, is shown in Fig. 51 for different values of the cross-correlation coefficient R and for $\varphi = 0$.

We note that the parameter φ, defined by the relation (40.14), characterizes the mean phase difference of the reverberation processes. In fact, if the following is computed with regard from (40.17):

$$\langle\Delta\psi\rangle = \int\limits_{-\pi}^\pi \Delta\psi W(\Delta\psi)\,d(\Delta\psi),$$

one is readily convinced that

$$\langle\Delta\psi\rangle = \varphi. \tag{40.18}$$

The variance of the fluctuations $\sigma^2_{\Delta\psi}$ is expressed for arbitrary values of R by very complex relations, but some notion may be gained as to its magnitude from the form of the distribution $W(\Delta\psi)$ shown in Fig. 51. With an increase in the correlation, generally speaking, the variance decreases, varying between the limits from $\pi^2/3$ (at R = 0) to zero (at R = 1).

We point out, in conclusion, that if the expression (40.15) is integrated over all possible values of $\Delta\psi$, we find the two-dimensional distribution $W(E_1, E_2)$ of the envelope of the reverberation signals, thus making it possible to determine, in particular, the cross-correlation function linking the values of $E_1(t)$ and $E_2(t)$ as a function of the cross-correlation of the initial reverberation signals.

This distribution is written as follows:

$$W\left(E_1, E_2\right) = \frac{E_1 E_2}{\sigma_{V_1}^2 \sigma_{V_2}^2 (1 - R^2 - S^2)} \exp\left[-\frac{E_1^2 + E_2^2}{2\sigma_{V_1}\sigma_{V_2}(1 - R^2 - S^2)}\right] I_0\left[\frac{E_1 E_2 (R^2 + S^2)^{1/2}}{\sigma_{V_1}\sigma_{V_2}(1 - R^2 - S^2)}\right] \qquad (40.19)$$

and is known as the t w o - d i m e n s i o n a l R a y l e i g h p r o b a b i l i t y d e n s i t y (see [1,12]).

§ 41. T r a n s m i s s i o n o f R e v e r b e r a t i o n T h r o u g h L i n e a r S y s t e m s

We now examine the energy characteristics of reverberation signals after their transmission through linear systems. This problem arises, for example, in the analysis of reverberation and its effects on various measuring instruments, in the filtering of reverberation and its envelope, and in certain other cases.

Let there be given a linear system having a complex transfer constant $K(\omega)$. If the energy spectrum of a stochastic process at the input to this system is equal to $G(\omega)$, the spectrum $G_1(\omega)$ and the correlation function $B_1(\tau)$ of the process at its output are determined by the relations

$$G_1\left(\omega\right) = \left|K\left(\omega\right)\right|^2 G\left(\omega\right), \qquad (41.1)$$

$$B_1\left(\tau\right) = (1/2\pi) \int_0^\infty \left|K\left(\omega\right)\right|^2 G\left(\omega\right) \cos \omega\tau \, d\omega. \qquad (41.2)$$

The value of the variance σ_1^2 at the output of the system coincides with the value of the correlation function (41.2) at $\tau = 0$ and may therefore be written in the form

$$\sigma_1^2 = (1/2\pi) \int_0^\infty \left|K\left(\omega\right)\right|^2 G\left(\omega\right) d\omega. \qquad (41.3)$$

The relations (41.1)-(41.3) enable one to compute the statistical characteristics of reverberation signals at the output of various types of linear systems from the known energy spectrum $G(\omega)$. These relations may be written in another form, if $G_1(\omega)$ and $B_1(\tau)$ are expressed in terms of the characteristics of the transmitted signals.

Recognizing that the reverberation spectrum is defined as (see Sec. 33)

$$G_{\mathbf{r}}\left(\omega\right) = 2\left\langle n_1\right\rangle \left\langle a^2\right\rangle \left|g_s\left(\omega\right)\right|^2,$$

where $g_s(\omega)$ is the amplitude spectrum of the transmitted signal, we go from (41.1)-(41.3) to the following relations:

$$G_{\mathbf{r}1}(\omega) = 2\left\langle n_1\right\rangle \left\langle a^2\right\rangle \left|K\left(\omega\right)\right|^2 \left|g_s\left(\omega\right)\right|^2, \qquad (41.4)$$

$$B_{\mathbf{r}1}\left(\omega\right) = (2\left\langle n_1\right\rangle \left\langle a^2\right\rangle / 2\pi) \int_0^\infty \left|K\left(\omega\right)\right|^2 \left|g_s\left(\omega\right)\right|^2 \cos \omega\tau \, d\omega, \qquad (41.5)$$

$$\sigma_{\mathbf{r}1}^2 = (2\left\langle n_1\right\rangle \left\langle a^2\right\rangle / 2\pi) \int_0^\infty \left|K\left(\omega\right)\right|^2 \left|g_s\left(\omega\right)\right|^2 d\omega. \qquad (41.6)$$

Let us consider a few examples of the application of the relations just obtained for calculation of the characteristics of reverberation signals at the output of filters with various transfer constants in the transmission of pulses with a sinusoidal carrier. It is assumed here that the reverberation is stationarized.

Let us suppose that the normalized transfer constant of the linear system is described by the function

$$K(\omega) = \exp[-(\omega - \omega_f)^2/\Delta\omega_f^2], \tag{41.7}$$

where ω_f is the tuning frequency of the filter, $\Delta\omega_f$ is its transmission band. We compute the parameters $G_{r1}(\omega)$, $B_{r1}(\tau)$, and σ_{r1}^2 at the output of this filter when it is known that its input is reverberation-generated in the radiation of pulses with a bell-shaped envelope.

Making use of the data of Table 11.1 (row 2), as well as Eqs. (41.4) and (41.7) for the reverberation spectrum, we obtain

$$G_{r1}(\omega) = \langle n_1 \rangle \langle a^2 \rangle \delta_{ef}^2 \exp(-A\omega^2 + B\omega - C), \tag{41.8}$$

where

$$\left.\begin{aligned}
A &= \frac{2\pi + (\Delta\omega_f \delta_{ef})^2}{2\Delta\omega_f^2}, \\[2mm]
B &= -\frac{4\pi\omega_f + 2\omega_0(\Delta\omega_f \delta_{ef})^2}{\pi\Delta\omega_f^2}, \\[2mm]
C &= \frac{2\pi\Delta\omega_f^2 + (\omega_0\Delta\omega_f \delta_{ef})^2}{\pi\Delta\omega_f^2}.
\end{aligned}\right\} \tag{41.9}$$

For the correlation function $B_{r1}(\tau)$ we obtain, on the basis of (41.5) and (41.8),

$$\begin{aligned}
B_{r1}(\tau) = (1/2\sqrt{\pi A}) \langle n_1 \rangle \langle a^2 \rangle \delta_{ef}^2 \exp(-C) \times \\
\times \{\exp[(B - j\tau)/A][1 - \Phi((B - j\tau)/2\sqrt{A})] + \\
+ \exp[(B + j\tau)^2/4A][1 - \Phi((B + j\tau)/2\sqrt{A})]\}.
\end{aligned} \tag{41.10}$$

The variance σ_{r1}^2 is determined from (41.10) if we let $\tau = 0$:

$$\sigma_{r1}^2 = (1/2\sqrt{\pi A}) \langle n_1 \rangle \langle a^2 \rangle \delta_{ef}^2 \exp(-C) \exp(B^2/4A)[1 - \Phi(B/2\sqrt{A})]. \tag{41.11}$$

We denote by

$$q^2 = \sigma_{r1}^2/\sigma_r^2$$

the ratio of the variances of the reverberation at the output and input, respectively, of the linear system and investigate the influence of the following two factors on the value of q^2: the broadening of the filter band for $\Delta\omega_f = \omega_0$, and the separation of the middle frequency of the reverberation spectrum from the center frequency of the filter at $\Delta\omega_f = \Delta\omega_{ef}$ (where $\Delta\omega_{ef}$ is the effective width of the reverberation spectrum).

The parameter σ_r^2 is determined from (41.11) by assuming that $\omega_f = \omega_0$ and passing to the limit as $\Delta\omega_f \to \infty$; we then obtain

$$\sigma_r^2 = \langle n_1 \rangle \langle a^2 \rangle \delta_{ef}/4. \tag{41.12}$$

Taking (41.11) and (41.12) into account, we obtain in the first case

$$q_1^2 = \frac{\sqrt{2\pi} \Delta F_f \delta_{ef}}{\sqrt{1 + 2\pi(\Delta F_f \delta_{ef})^2}}; \tag{41.13}$$

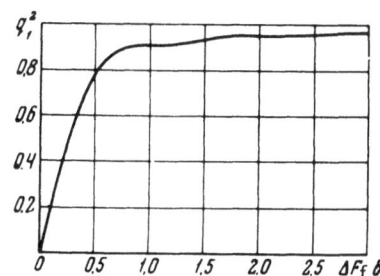

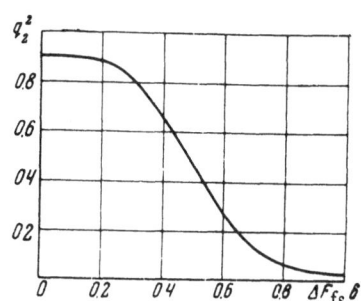

Fig. 52. Dependence of the normalized variance of reverberation on the relative bandwidth of the filter.

Fig. 53. Dependence of the normalized variance of reverberation on the relative separation of the center frequency of the filter.

in the second case,

$$q_2^2 = [\sqrt{\pi} / 2 (2 + \pi)] \exp [- 8\pi (\Delta F_{fs} \ \delta_{ef})^2 / 2 + \pi], \tag{41.14}$$

where

$$2\pi \Delta F_{fs} = \omega_s - \omega_0, \quad \Delta \omega_f = 2\pi \Delta F_f .$$

If a rectangular pulse of duration δ is radiated, and the transfer constant $K(\omega)$ is determined by the relation (41.7), then, making use of the data of Table 11.1 (row 1), as well as Eq. (41.4), we obtain for q_1^2 and q_2^2, respectively,

$$q_1^2 = (^1/_2) [1 + \Phi (\pi \Delta F_f \delta_{ef} / \sqrt{2}) - (\Delta F_f \delta_{ef} / \sqrt{2\pi^3}) \exp (-\pi^2 (\Delta F_f \delta_{ef})^2)], \tag{41.15}$$

$$q_2^2 = \exp [- 2 (\Delta F_{fs} \ \delta_{ef})^2] \left\{ \left[1 + \Phi \left(\frac{\pi}{\sqrt{2}} + j\Delta F_{fs}\delta_{ef} \right) \right] \times \right.$$
$$\left. \times [1 + j \sqrt{2}\Delta F_{fs} \ \delta_{ef} / \pi] - (1 / \sqrt{\pi^3}) \exp [-(\pi + j \sqrt{2}\Delta F_{fs} \ \delta_{ef})^2] \right\} . \tag{41.16}$$

Let us consider another sample calculation of the parameters q_1^2 and q_2^2, when the transfer constant of the linear system is described by the plateau characteristic

$$K (\omega) = \begin{cases} 1 \text{ for } \omega_1 - \Delta\omega_f / 2 \leqslant \omega < \omega_1 + \Delta\omega_f / 2, \\ 0 \text{ outside the interval,} \end{cases} \tag{41.17}$$

where ω_1 and $\Delta\omega_f$ are the center frequency and bandwidth of the filter.

For the radiation of a bell-shaped pulse we arrive at the relations

$$q_1^2 = \Phi (\sqrt{\pi}\Delta F_f \delta_{ef}), \tag{41.18}$$

$$q_2^2 = (^1/_2) [\Phi (2 \sqrt{\pi}\Delta F_{fs} \ \delta_{ef} + \sqrt{\pi} / 2) - \Phi (2 \sqrt{\pi}\Delta F_{fs} \ \delta_{ef} - \sqrt{\pi} / 2)] . \tag{41.19}$$

Finally, in the case of radiation of a rectangular pulse, we have for the transfer constant $K(\omega)$ defined according to (41.17),

$$q_1^2 = (2 / \pi) \left[\mathrm{Si} (\pi \Delta F_f \ \delta_{ef})^2 - \frac{\sin^2 (\pi \Delta F_f \delta_{ef} / 2)}{\pi \Delta F_f \ \delta_{ef} / 2} \right], \tag{41.20}$$

$$q_2^2 = (1/\pi)\left[\text{Si}\,(4\pi\Delta F_{fs}\,\delta_{ef}+2\pi)-\text{Si}\,(4\pi\Delta F_{fs}\,\delta_{ef}-2\pi)-\right.$$
$$\left.-\frac{\sin^2(2\pi\Delta F_{fs}\,\delta_{ef}+\pi)}{2\pi\Delta F_{fs}\,\delta_{ef}+\pi}+\frac{\sin^2(2\pi\Delta F_{fs}\,\delta_{ef}-\pi)}{2\pi\Delta F_{fs}\,\delta_{ef}-\pi}\right].\tag{41.21}$$

An analysis of the resultant relations shows that the reverberation level at the output of the linear system increases, tending to a definite limit as the transmission band of the filter is expanded, and falls off, tending to zero as the difference between the tuning frequency of the filter and the center frequency of the reverberation spectrum is increased.

This is illustrated graphically in Figs. 52 and 53, which show the functions $q_1^2 = f(\Delta F_f \delta_{ef})$ and $q_2^2 = f(\Delta F_{fs} \delta_{ef})$, plotted in accordance with Eqs. (41.20) and (41.21).

We note that the relations (41.4)-(41.6) may also be used analogously to calculate other statistical characteristics of reverberation signals at the output of particular linear systems.

We next consider the transmission of fluctuations of the reverberation envelope through linear systems.

It proves inappropriate to calculate the correlation function $B_{rE1}(\tau)$, the variance σ_{E1}^2, and the energy spectrum $G_{rE1}(\omega)$ at the output of a linear system with transfer constant $K(\omega)$ directly from Eqs. (41.1)-(41.3) in the given case.

The final results are more speedily obtained by introducing the so-called transfer function

$$H(\tau) = (1/\pi)\int_0^\infty |K(\omega)|^2 \cos\omega\tau\,d\omega.\tag{41.22}$$

Making use of (41.22), the following relations may be written; they are obtained from (41.1)-(41.3) by applying the convolution theorem:

$$B_{rE1}(\tau) = 2\int_0^\infty H(t)\,B_{rE}(\tau-t)\,dt,\tag{41.23}$$

$$\sigma_{E1}^2 = 2\int_0^\infty H(t)\,B_{rE}(t)\,dt,\tag{41.24}$$

$$G_{rE1}(\omega) = 2\int_0^\infty B_{rE1}(\tau)\cos\omega\tau\,d\tau.\tag{41.25}$$

The relevant parameters, therefore, are expressed solely in terms of the correlation function of the reverberation envelope fluctuations, which is fairly easily calculated according to the characteristics of the radiated signals (see Sec. 24).

As an example of the application of the derived relations, we consider the transmission of the reverberation envelope through a low-frequency filter with transfer constant

$$K(\omega) = \exp\left[-(\omega/\Delta\omega_f)^2\right].$$

Taking (41.22) into account, we write the following expression for the transfer function of the given filter:

$$H(\tau) = (\tfrac{1}{2}\sqrt{2\pi})\,\Delta\omega_f\exp\left[-(\Delta\omega_f\tau)^2/8\right].\tag{41.26}$$

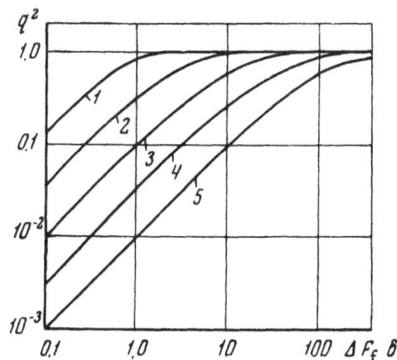

Fig. 54. Dependence of the normalized variance of reverberation envelope fluctuations on the relative bandwidth of a low-frequency filter for the transmission of noisy signals. 1) $\Delta F_{ef}\delta = 1$; 2) 3; 3) 10; 4) 30; 5) 100.

Let us assume that the radiated signals represent noisy pulses of duration δ. Then, for the correlation function $B_{rE}(\tau)$ of the reverberation envelope fluctuations the following relation is valid (see Sec. 25):

$$B_{rE}(\tau) \approx \sigma_E^2 r_x^2(\tau)(1 - |\tau|/\delta)^2, \quad |\tau| \leqslant \delta, \quad (41.27)$$

where $r_x(\tau)$ is the envelope of the correlation coefficient of the noise carrier. If we assume further that the filter forming this carrier has a bell-shaped characteristic, we have for $r_x(\tau)$

$$r_x(\tau) \approx \exp[-(\Delta\omega_{ef}\tau)^2/4\pi],$$

where $\Delta\omega_{ef}$ is the effective bandwidth of the high-frequency filter. The correlation function (41.27) in this case assumes the form

$$B_{rE}(\tau) \approx \sigma_E \exp[-(\Delta\omega_{ef}\tau)^2/2\pi](1 - |\tau|/\delta)^2. \quad (41.28)$$

Making use of (41.24), (41.26), and (41.28), we calculate the variance σ_{E1}^2 at the output of the linear system.

Introducing the parameter

$$q^2 = \sigma_{E1}^2/\sigma_E^2,$$

the calculation yields the following result:

$$q^2 \approx \frac{2\Omega}{\sqrt{\pi\Omega^2 + 4k^2}} \left\{ (\sqrt{\pi}/2)\Phi\left[(\pi/2)^{1/2}\sqrt{\pi\Omega^2 + 4k^2}\right]\left(1 + \frac{1}{\sqrt{\pi\Omega^2 + 4k^2}}\right) + \frac{\exp[-(\pi/2)(\pi\Omega^2 + 4k^2) - 2]}{\sqrt{2\pi}\sqrt{\pi\Omega^2 + 4k^2}} \right\}, \quad (41.29)$$

where

$$\Omega = \Delta F_f\delta, \quad k = \Delta F_{ef}\delta,$$

$$\Delta\omega_f = 2\pi\Delta F_f, \quad \Delta\omega_{ef} = 2\pi\Delta F_{ef}.$$

A family of curves for $q^2(\Delta F_f\delta)$, constructed for various values of the parameter $\Delta F_{ef}\delta$, is shown in Fig. 54.

An analysis of the final relation (41.29) leads to the following conclusion with regard to the dependence of the variance of the reverberation envelope fluctuations on the parameters of the linear system:

The variance increases, tending to a definite limit* as the bandwidth of the filter is increased.

For a constant bandwidth on the part of the low-frequency filter and expansion of the band of the filter generating the noise carrier, the variance decreases, tending to zero.

---

*This limit corresponds to the variance for fluctuations obeying a Rayleigh distribution law.

§42. Differentiation of Reverberation and Its Envelope

The investigation of the statistical characteristics of the derivative of reverberation and its envelope, to which the present section is devoted, is important in connection with the analysis of excessive peak levels that over-exceed some specified level, the investigation of reverberation transmission through certain types of linear and nonlinear devices, and the solution of other problems.

We start with some important definitions.

The derivative $\hat{V}(t)$ of a stochastic process V(t) at the time t refers to a stochastic process defined as (see [1,12,19])

$$\dot{V}(t) = dV(t)/dt = \lim_{\Delta t \to 0} \frac{V(t+\Delta t) - V(t)}{\Delta t}.$$

(42.1)

This expression is normally interpreted as convergence in the mean square, as depicted by the limit

$$\lim_{\Delta t \to 0} \left\langle \left| \dot{V}(t) - \frac{V(t+\Delta t) - V(t)}{\Delta t} \right| \right\rangle = 0.$$

The process $\hat{V}(t)$ generated according to (42.1) describes the rate of change of V(t) at any instant of time.

As before, we assume that the reverberation is stationarized, and we are concerned with the derivative of a process that is stationary at least in the broad sense. This permits the following interesting property to be formulated at once: The reverberation is not correlated with its derivative, i.e.,

$$\langle V(t) \dot{V}(t) \rangle = 0.$$

(42.2)

The truth of the property (42.2), which is general for stationary stochastic processes, can be demonstrated, for example, by differentiation of the reverberation correlation function.

In fact, taking into account the definition of the correlation function

$$B_{\mathrm{r}}(\tau) = \langle V(t) V(t+\tau) \rangle,$$

taking the derivative of the right- and left-hand sides with respect to t, and recognizing the fact that for stationary stochastic functions describing physical processes, for any τ,

$$dB_{\mathrm{r}}(\tau)/dt = 0,$$

we obtain*

$$\langle \dot{V}(t) V(t+\tau) \rangle + \langle V(t) \dot{V}(t+\tau) \rangle = 0.$$

If now we let $\tau = 0$, we arrive at Eq. (42.2).

In the case when $\tau \neq 0$, it is necessary, in order to evaluate the cross-correlation functions $\langle \hat{V}(t)V(t+\tau)\rangle$ and $\langle V(t)\hat{V}(t+\tau)\rangle$, in general, to know the joint probability density $W_V(V, V_\tau)$. For the calculation of the cross-correlation functions of reverberation, however, if the latter is stationary, it is admissible to use the general relation

---

*The given process must, of course, be such as to permit the order of statistical averaging and differentiation to be interchanged.

$$dB_\Gamma(\tau)/d\tau = \langle V(t)\dot{V}(t+\tau)\rangle, \tag{42.3}$$

which is valid as long as the derivative of the reverberation correlation function exists; in the transmission of physically realizable signals this condition is clearly fulfilled for reverberation.

For investigations of the statistical characteristics of the derivative of reverberation $\dot{V}(t)$, it is advantageous to know at least its one-dimensional probability density $W(\dot{V})$, which can be obtained from the two-dimensional distribution $W_V(V, V_\tau)$ and the relation (42.1).

Performing a substitution of variables in $W_V(V, V_\tau)$ according to the formulas

$$V = V, \ V_\tau = V + \Delta t\,(\Delta V/\Delta t),$$

where

$$\Delta V(t) = V(t + \Delta t) - V(t),$$

it can be shown that $W(\dot{V})$ is expressed in terms of $W_V(V, V_\tau)$ in the following manner:

$$W(\dot{V}) = \int_{-\infty}^{\infty} \lim_{\Delta t \to 0} \{\Delta t W_V[V, V + \Delta t\,(\Delta V/\Delta t)]\}\, dV, \tag{42.4}$$

where

$$\lim_{\Delta t \to 0}(\Delta V/\Delta t) = \dot{V}.$$

If we account for the fact that the joint distribution $W_V(V, V_\tau)$ of the instantaneous values of the reverberation is normal, i.e.,

$$W_V(V, V_\tau) = \frac{1}{2\pi\sigma_V^2\,[1 - R_\Gamma^2(\tau)]^{1/2}}\exp\left\{-\frac{V^2 + V_\tau^2 - 2VV_\tau R_\Gamma(\tau)}{2\sigma_V^2\,[1 - R_\Gamma^2(\tau)]}\right\}, \tag{42.5}$$

then substitution of Eq. (42.5) into (42.4) yields

$$W(\dot{V}) = \frac{1}{[-2\pi\sigma_V^2 R_\Gamma''(0)]^{1/2}}\exp\left[-\frac{\dot{V}^2}{-2\sigma_V^2 R_\Gamma''(0)}\right], \tag{42.6}$$

where

$$R_\Gamma''(0) = d^2 R_\Gamma(\tau)/d\tau^2\,|_{\tau=0}. \tag{42.7}$$

It is apparent from (42.6) that the probability density found for the derivative of reverberation corresponds to a normal distribution with variance

$$\sigma_{\dot{V}}^2 = -\sigma_V^2 R_\Gamma''(0). \tag{42.8}$$

We note that the variance of the derivative, provided the parameter $R_\Gamma''(0)$ exists, is calculated by means of the relation (42.8) for any distribution of the original stochastic process.

On the assumption of a stationary process $\dot{V}(t)$ and validity of (42.6), it may be assumed that the n-dimensional distribution $W(\dot{V}_1, \dot{V}_2, \ldots, \dot{V}_n)$ obeys a normal law with the elements of the correlation matrix defined by the relation

$$R(\tau) = \frac{d^2 B_\Gamma(\tau)/d\tau^2}{d^2 B_\Gamma(\tau)/d\tau^2\,|_{\tau=0}}. \tag{42.9}$$

The results obtained here permit the solution of such problems as the statistical characteristics of the instantaneous reverberation periods, transitions through zero, the energy of the peaks between neighboring intersections, etc. The methods of solving these problems have been well developed in the theory of stochastic functions and may be found, for example, in [1, 12, 19, 51].

We examine next some of the statistical characteristics of the derivative of the reverberation envelope.

Proceeding on the basis of the two-dimensional Rayleigh distribution of the envelope (40.19), we find the joint probability density $W(E, \dot{E})$, making use of the general relation

$$W(E, \dot{E}) = \lim_{\Delta t \to 0} \{\Delta t W_E [E, E + \Delta t (\Delta E/\Delta t)]\}, \tag{42.10}$$

where

$$\lim_{\Delta t \to 0} (\Delta E/\Delta t) = \dot{E}.$$

This distribution has the form

$$W(E, \dot{E}) = \frac{E}{\sqrt{2\pi}\sigma_r^3 \Delta\Omega} \exp\left(-\frac{E^2 + \dot{E}^2/\Delta\Omega^2}{2\sigma_V^2}\right), \tag{42.11}$$

where $\Delta\Omega$ is the mean-square frequency of the spectrum of the reverberation envelope, and is defined as

$$\Delta\Omega = [\Omega_2 - \Omega_1^2]^{1/2}, \tag{42.12}$$

where the parameters

$$\Omega_k = \frac{\int\limits_{-\infty}^{\infty} \omega^k G_{rE}(\omega)\, d\omega}{\int\limits_{-\infty}^{\infty} G_{rE}(\omega)\, d\omega} \tag{42.13}$$

are computed as the moments of the energy spectrum $G_{rE}(\omega)$ of the reverberation envelope fluctuations.

The one-dimensional distribution of the derivative

$$W(\dot{E}) = \int\limits_0^{\infty} W(E, \dot{E})\, dE,$$

calculated by means of (42.11), turns out to be equal to*

$$W(\dot{E}) = \frac{1}{\sqrt{2\pi}\sigma_V \Delta\Omega} \exp\left(-\dot{E}^2/2\sigma_V^2 \Delta\Omega^2\right). \tag{42.14}$$

Hence, it follows that <u>the derivative of the reverberation envelope is distributed according to a normal law</u>, with variance

$$\sigma_{\dot{E}}^2 = \sigma_V^2 \Delta\Omega^2. \tag{42.15}$$

If we bear in mind that the distribution of the reverberation envelope is described by a Rayleigh law, then, comparing (18.2), (42.11), and (42.14), we arrive at the relation

---

* The details of the derivation of Eqs. (42.11) and (42.14) are given, for example, in [1].

$$W(E, \dot{E}) = W(E) \, W(\dot{E}), \tag{42.16}$$

which implies that the reverberation envelope and its derivative are statistically independent.

As an example we find the one-dimensional probability density function $W(\dot{V})$ of the derivative of reverberation for the transmission of a bell-shaped pulse with sinusoidal carrier. Making use of the expression for the reverberation correlation coefficient (see Table 22.1), we find
on the basis of (42.7),

$$R_{\mathrm{r}}''(0) = - [(\pi/2\delta_{\mathrm{ef}}^2) + \omega_0^2]. \tag{42.17}$$

Then, taking (42.6) into account, we obtain, for the sought-after distribution,

$$W(\dot{V}) = \frac{1}{\sqrt{2\pi \, [(\pi/2\delta_{\mathrm{ef}}^2) + \omega_0^2] \, \sigma_V}} \exp\{- \dot{V}^2/2\sigma_V^2 \, [(\pi/2\delta_{\mathrm{ef}}^2) + \omega_0^2]\}. \tag{42.18}$$

§ 43. Stationarization of Reverberation Signals and Characteristics of Reverberation as a Nonstationary Process

One of the important postulates that has frequently recurred in the foregoing discussions
is contained in the following statement: Reverberation represents a process reducible to one that is stationary.

This assumption is applicable in the case when, over an interval $(t - T/2, t + T/2)$ no
smaller than the effective duration of the transmitted signals:

$$T > \delta_{\mathrm{ef}}, \tag{43.1}$$

the variation of the average number of elementary scattered signals observed at the point of reception and of the mean-square amplitude of these signals is negligibly small. Such conditions,
as a rule, are observed for running instants of time t satisfying the inequality

$$t \gg \delta_{\mathrm{ef}}. \tag{43.2}$$

The variations of the mean reverberation intensity are inconsequential on the indicated intervals, i.e.,

$$\langle F^2(t - T/2) \rangle \approx \langle F^2(t + T/2) \rangle. \tag{43.3}$$

For the experimental investigation of the statistical properties of reverberation it is desirable in most instances to have fairly long stationary reverberation lengths available. Such
lengths are obtainable in practice for times satisfying the condition (43.2).

In this connection, we pause to consider the methods by which the stationarization of reverberation signals was realized in the measurements and processing of the experimental data.

Two methods of stationarization were used, viz.:

Time-based automatic gain control (TAGC) according to some preselected law;

Automatic gain control (AGC) based on the averaged reverberation level.

In the first case, the gain K(t) of the receiving section was varied for each reverberation
state, beginning with the instant of signal transmission, according to the law

$$K(t) = K_{\max} (t/t_{\max})^A 10^{B(t-t_{\max})},$$ (43.4)

where $K_{\max}$ is the maximum gain at the time $t_{\max}$; A and B are parameters which could be varied between the required limits.

The law (43.4), given a proper choice of the parameters A and B, ensures compensation of the time variation in reverberation level due to spreading of the sound rays and absorption in the ocean.

As a rule, the TAGC system was used directly for ocean measurements and was incorporated into the receiving instrumentation complex (see Sec. 45). It was possible by means of TAGC to stationarize the reverberation signals over intervals of 0.2-0.4 sec for transmitted pulses with a duration of no more than 10 msec and on intervals of 2-3 sec for longer signal durations. It was not feasible to stationarize to any great extent reverberation signals produced by combined surface and bottom scattering, because the law of decay of the mean reverberation level in this case was described by functions more complex than $1/K(t)$, where $K(t)$ is defined as in (43.4).

The second method of stationarization consists in the following. A signal was fed to the input of the AGC network with its level proportional to the reverberation envelope, averaged over some interval T_a. This interval was chosen so as to obtain the largest stationarized reverberation length with minimum distortion of its stationary component.

An investigation of the characteristics of stationarization by means of AGC showed that the most favorable ratio T_a/δ_{ef} lies between the limits

$$T_a/\delta_{ef} = 3 \div 5.$$ (43.5)

The methods used in the measurements for stationarization of the reverberation signals made it possible to obtain stationary intervals equal to 15-30 times δ_{ef}, which meant that the experimental data could be processed with statistical averaging over the time.

The investigation of reverberation at times t not satisfying the conditions (43.2) shows that in this case the reverberation cannot be stationarized, and that its statistical characteristics will necessarily differ considerably from the corresponding characteristics determined when the indicated condition is fulfilled.

Examples of oscillograms of near-zone reverberation with nonstationary intervals are shown in Fig. 55, from which it is apparent that for near reverberation the correlation interval of its envelope fluctuations is smaller than for the farther reverberation zone. For this reason, it is instructive to investigate certain properties of reverberation, taking into account the influence of nonstationarity.

We find an expression for the semi-invariants of the two-dimensional distribution, assuming that at any time t every state of the reverberation signal $F_n(t)$ is represented in the form

$$F_n(t) = \sum_{i=1}^{n} \alpha_i f(t_i) s(t - t_i),$$ (43.6)

where

$$a(t_i) = \alpha_i f(t_i)$$ (43.7)

is the amplitude of the i-th elementary scattered signal at the point of reception, α_i is a coefficient characterizing the scattering power of the i-th inhomogeneity, $f(t)$ is a function accounting for the attenuation of the signal due to divergence of the sound rays and absorption.

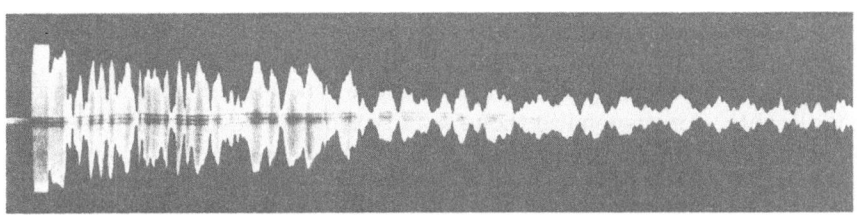

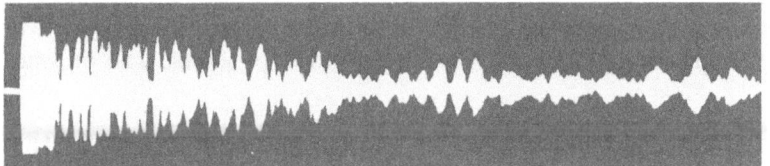

Fig. 55. Oscillograms of near-zone reverberation.

Applying the two-dimensional theorem of superposition of stochastic perturbations to $F_n(t)$ (see Sec. 6), we obtain for the $(k + l)$-th order semi-invariants

$$\lambda_{kl}(t_1, t_2) = \langle\alpha^{k+l}\rangle \int_{-\infty}^{\infty} \langle n_1\rangle(t')\rangle f^{k+l}(t') s^k(t_1 - t') s^l(t_2 - t') dt', \qquad (43.8)$$

where the parameter $\langle\alpha^{k+l}\rangle$ is defined analogously with (6.3) in terms of the probability density $W(\alpha)$ determined on the assumption of statistical uniformity of the spatial scattering characteristics, which are independent of the running time t; the parameter $\langle n_1(t)\rangle$ is the average number of elementary scattered signals observed at the reception point per unit time.

The reverberation correlation function is obtainable from (43.8) and will, by virtue of the nonstationarity of the investigated process, depend on the time.

Performing the substitution of variables

$$t_1 = t, \quad t_2 = t + \tau$$

in (43.7) and letting k = 1, l = 1, we find

$$B_r(\tau, t) = \langle\alpha^2\rangle \int_{-\infty}^{\infty} \langle n_1(t')\rangle f^2(t') s(t - t') s(t + \tau - t') dt'. \qquad (43.9)$$

It is interesting to note that for instants of time when the following approximate equations are satisfied:

$$\left.\begin{array}{l} \langle n_1(t - \delta_{ef}/2)\rangle \approx \langle n_1(t + \delta_{ef}/2), \\ f^2(t - \delta_{ef}/2) \approx f^2(t + \delta_{ef}/2), \end{array}\right\} \qquad (43.10)$$

i.e, when the functions $\langle n_1(t)\rangle$ and $f^2(t)$ are smoothly varying on the interval $(t - \delta_{ef}/2, t + \delta_{ef}/2)$, the relation (43.8) may be approximately written in the form

$$\lambda_{kl}(t) \approx \langle\alpha^{k+l}\rangle f^{k+l}(t) \langle n_1(t)\rangle \int_{-\infty}^{\infty} s_1^k(t_1 - t') s_2^l(t_2 - t') dt'. \qquad (43.11)$$

If, in addition, we consider, in correspondence with (43.7), that

$$\langle a^{k+l}(t) \rangle = \langle \alpha^{k+l} \rangle f^{k+l}(t),$$

then (43.11) goes over to the previous relation (6.17).

The time intervals under which conditions of the type (43.10) prevail, to the extent that Eq. (43.11) is applicable, correspond to those sections of the reverberation process where it can be stationarized. For the functions $<n_1(t)>$ and $f(t)$ normally realized in practice, the conditions (43.10) are equivalent to the inequality (43.2) given above.

The general relation (43.8) may be used also to find such statistical characteristics of reverberation as the variance, coefficient of excess, energy spectrum, etc. These characteristics, because of the nonstationarity of the investigated process, will depend, like λ_{kl}, on the time.

As an example we find the reverberation correlation coefficient for the transmission of a pulse with a sinusoidal carrier and rectangular envelope (10.7). Assuming that the scattering of sound occurs at inhomogeneities concentrated in a layer (for example, the case of surface reverberation), we write for the factor $<n_1(t)> f^2(t)$ in front of the integral sign in Eq. (43.8),

$$\langle n_1(t) \rangle f^2(t) = \langle n_1(t_0) \rangle f^2(t_0)(t_0/t)^3, \quad t \geqslant t_0, \; t_0 > 0, \tag{43.12}$$

where t_0 is some instant of time, after which the law (43.12) holds true.

Taking (10.6), (10.7), (43.9), and (43.12) into account, carrying out the integration, and normalizing for the envelope of the correlation coefficient, we obtain

$$r_{\mathrm{r}}(\tau, t) = \frac{\{(t/\delta)^2 - [(t/\delta) - (1 - |\tau|/\delta)]^2\} [(t/\delta) - 1]^2}{[(t/\delta) - (1 - |\tau|/\delta)]^2 \{(t/\delta)^2 - [(t/\delta) - 1]^2\}}. \tag{43.13}$$

Figure 56 shows a graph of a family of functions $r_{\mathrm{r}}(\tau)$ constructed for various values of the parameter t/δ characterizing the relative time measured from transmission of the pulse. It is evident from this graph that the smaller the parameter t/δ, the narrower will be the time correlation of the reverberation.

On the other hand, as the reading time is increased, the dependence of $r_{\mathrm{r}}(\tau)$ on the parameter t/δ attenuates, and already for

$$(t/\delta) > 10$$

better agreement is observed between the correlation coefficient (43.13) and the value calculated previously for the case of stationarized reverberation (Table 21.1, row 1). This conclusion is qualitatively confirmed by the form of the near-reverberation envelope whose oscillograms are presented in Fig. 55. In fact, as implied by Eqs. (43.13) and Fig. 56, for the instants of time immediately following the transmitted signal the correlation interval turns out to be less than for later times; the oscillograms presented in Fig. 55 show that the envelope of near reverberation does indeed fluctuate somewhat more rapidly.

Similar conclusions are inferred from [35], which is devoted to an investigation of a reverberation model on the assumption that the distribution of scatterers in the ocean obeys a Poisson law, while the reverberation represents a nonstationary stochastic process.

§ 44. Reverberation during Continuous Transmission

Certain statistical characteristics of reverberation are conveniently studied using the transmission of continuous signals with one kind of modulation or another. In this case, the time fluctuations of the reverberation are caused by the form of the amplitude and frequency modulation of the signal, the motion of the scatterers in the ocean, the displacement of the

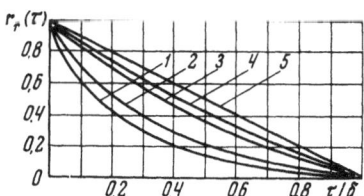

Fig. 56. Calculated values of the correlation coefficient envelope with allowance for nonstationarity of reverberation. 1) $t/\delta = 1.5$; 2) 2; 3) 3; 4) 10; 5) ∞.

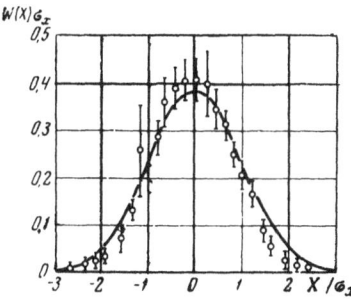

Fig. 57. Displacement distribution of the ocean surface. The circles correspond to the experimental data, the curve to a normal distribution.

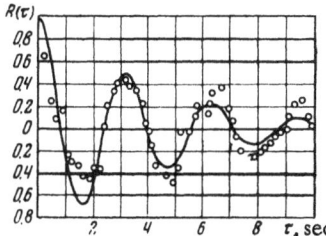

Fig. 58. Sample autocorrelation coefficient for the displacements of the ocean surface.

acoustic arrays, and the time variation of the scattered signal levels and conditions of sound propagation. Moreover, in the transmission, say, of frequency-modulated signals, the variations of the reverberation level may also be determined by the distribution of the scatterers in space.

Consequently, the investigation of the statistical properties of reverberation in the transmission of continuous signals can in some cases yield additional information regarding the model of reverberation as a stochastic process and explain the characteristics of continuous operation in sonar systems [30, 41-43, 53, 55].

In this section we give the results of our own research on surface reverberation in the transmission of continuous sinusoidal signals, also presenting a short review of the investigations mentioned above.

In studying surface reverberation, caused mainly by scattering at inhomogeneities of the surface layer, we simultaneously measured the probability distributions for the displacement of points on the surface and the time correlation of these displacements. The investigations were carried out by means of an electrical wavemeter during observation of the ocean surface at a single point.

We present here certain experimental data on the statistical properties of the ocean surface wave state.

A graph is presented in Fig. 57, showing the displacement distribution of surface points with a surface state of 2 to 4 points on the sea-height scale; the graph represents the results of ten series of measurements. It is evident from this figure that the distribution found from the experiments is close to a normal distribution but has positive excess and a slight asymmetry.

For a wave state exceeding 3-4 points, the distribution of the surface displacements retains its form in the majority of cases, but occasionally a distribution with two peaks is observed, the presence of which is clearly attributable to the simultaneous presence of swell and wind waves.

We next examine the results of measurements of the time correlation of the displacement of the ocean surface. These displacements have a quasi-harmonic character, as confirmed by the form of their autocorrelation coefficient $R(\tau)$ (Fig. 58). In many cases $R(\tau)$ is approximated with sufficient accuracy by a dependence of the form

$$R(\tau) = \exp(-\alpha|\tau|)\cos\Omega_0\tau, \qquad\qquad (44.1)$$

where the parameters α and Ω_0 are determined by the nature of the wave state. These parameters are readily found from the experimental data when one considers the fact that Ω_0 is related to the mean period of the waves $<T_w>$ by the expression

$$\Omega_0 \approx 2\pi/\langle T_w\rangle. \qquad\qquad (44.2)$$

Thus, for a wave state corresponding to the data of Fig. 58, we obtain

$$\alpha \approx 0.23 \text{ sec}^{-1}, \qquad \Omega_0 \approx 2 \text{ sec}^{-1}.$$

The given characteristics of the wave state permit certain assumptions to be made with respect to the motion of the scatterers in the surface layer.

If we assume that the scatterers are situated primarily in a layer up to 10-20 m thick just beneath the ocean surface, their motion ought to correlate with the displacement of the surface points, because we know that the motion in the main body of water due, for example, to a sea height of 2-3 points permeates to depths of 30-40 m or more. The motion of the scatterers, most of which are air bubbles, should follow nearly elliptical trajectories, with an average period approximately equal to $<T_w>$, due to the quasi-harmonic character of the surface wave state of the ocean. Also possible is translational motion of the bubbles in a horizontal plane, which could be elicited, in particular, by the ocean currents.

We now present the fundamental results of an analysis of the statistical characteristics of surface reverberation during the transmission of continuous sinusoidal signals and compare them with the analogous characteristics of the motion of the ocean surface.

In performing the ocean measurements, the effective scattering volume, which takes in part of the surface layer, was determined by intersecting the directivity patterns of the transmitting and receiving arrays; they turned out to be located at distances of 500-800 m from the arrays.

The following characteristics were ascertained as a result of the reverberation measurements:

The probability distribution of the envelope;

The correlation interval of the envelope and phase fluctuations;

The correlation functions of the envelope.

The investigations were conducted in a frequency range from 4 to 36 kc, enabling us in some cases to find the frequency dependence of the measured quantities.

Assuming a uniform distribution of the instantaneous phase fluctuations of the reverberation in the interval $(0, 2\pi)$, and a normal distribution of its instantaneous values, we at once deduce a Rayleigh law for the envelope. However, a uniform phase distribution will occur when the motion of the scatterers is such that the phases of the elementary scattered signals fluctuate at least in the interval $(0, 2\pi)$. The latter is true in the event that

$$(4\langle l_s\rangle/\lambda) > 1, \qquad\qquad (44.3)$$

where $<l_s>$ is the average displacement of the scatterers in the direction of the acoustic arrays; λ is the wavelength corresponding to the frequency of the transmitted signals.

If the inequality (44.3) is not satisfied, the phase distribution of the reverberation will not be uniform in the interval indicated, which means that the envelope distribution will deviate from a Rayleigh law.

The values calculated for the coefficient of variation of the reverberation envelope on the basis of the experimental data for various transmitted-signal frequencies are presented below.

f_0, kc	4	7	11	15	36
γ_v	0.21—0.25	0.33—0.35	0.4—0.41	0.4—0.42	0.45—0.47

The conditions under which this series of measurements were carried out were characterized by the following parameters of the ocean surface wave state:

$$\sigma_{\Delta h} \approx 15\,\text{cm}, \qquad \langle T_w \rangle \approx 8\,\text{sec},$$

where $\sigma_{\Delta h}$ is the rms sea height.

The quantity $\langle \Delta l_s \rangle$ and the rms velocity of the scatterers σ_s are defined as

$$\langle \Delta l_s \rangle = k_1 \sigma_{\Delta h}, \qquad \sigma_s \approx 4 \langle l_s \rangle / \langle T_w \rangle, \tag{44.4}$$

where k_1 is a coefficient relating the displacements of a point of the ocean surface to the motion of the scatterers; it assumes values ranging from 0 to 1.

In the present case, assuming, for example, $k_1 = 0.5$, we find, on the basis of (44.4),

$$\langle \Delta l_s \rangle \approx 7.5\,\text{cm}, \qquad \sigma_s \approx 3.7\,\text{cm/sec}.$$

Consequently, for frequencies below 7 kc, the inequality (44.3) is weakly satisfied, thus explaining the small values of the coefficient of variation for low frequencies. From the above data it also follows that as the frequency is increased, the coefficient of variation will increase, approaching a value of 0.52, which corresponds to a Rayleigh distribution.

We now present data on the influence of the sea height on the coefficient of variation of the envelope. These data are given in Table 44.1 for two states of the ocean surface.

It is evident from the table that the coefficient of variation tends to increase in general with increasing sea height and with increasing frequency.

The results of measurements performed with a sea height of less than one point in the absence of wind waves are presented below:

f_0, kc	4	7	11	15	25	36
γ_v	0.07	0.12	—	0.2	0.22	0.22
σ_ψ, rad	0.2—0.25	—	0.4	—	—	—

It turned out that for a frequency of 4 kc the coefficient of variation fell between the limits 0.05 and 0.1, while the reverberation phase fluctuated by about 0.23 rad. The coefficient of variation in this case increased approximately in proportion to the frequency up to 15 kc, after which the rate of increase eased off.

The cited experimental data on the parameters of the distributions of the reverberation envelope are consistent with the notions of the model of sound scattering by inhomogeneities of the near-surface layer, such as air bubbles, the motion of which is determined by the state of the ocean surface.

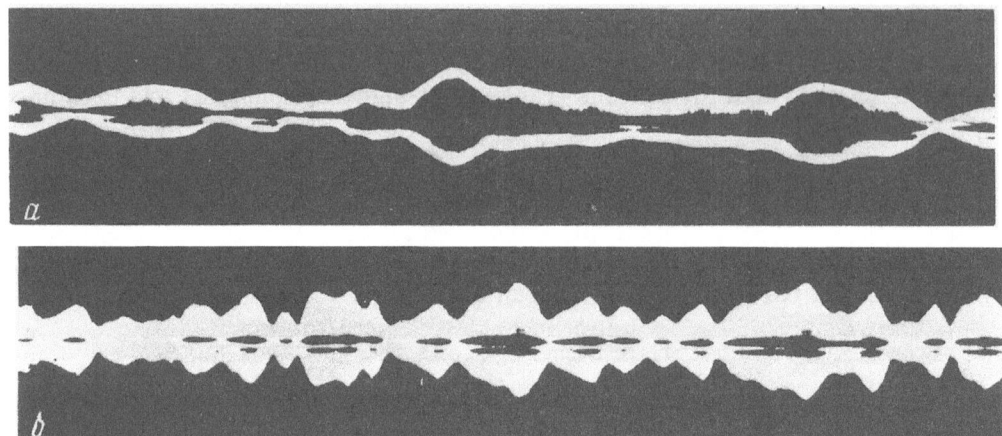

Fig. 59. Reverberation oscillograms obtained during the transmission of continuous sinusoidal signals of various frequencies. a) f_0 = 4 kc; b) 15 kc.

Table 44.1. Values of the Coefficient of Variation
of the Envelope for Different Sea States

Surface state	$\sigma_{\Delta h},$ cm	Envelope coefficient of variation γ_y					
		f_0, kc					
		4	7	11	15	25	36
Swell	10	0.22	0.35	—	—	0.47	0.49
	22	—	—	0.38	0.45	—	0.49
	31	0.25	—	0.4	0.47	—	0.5
Swell with wind waves	12	0.24	—	—	0.4	0.48	—
	26	—	0.38	—	0.41	—	0.5
	45	0.25	0.32	—	0.5	0.51	—

We proceed now to analyze the results of an investigation of the correlation characteristics of reverberation.

It follows from the general notions regarding the influence of scatterer motion on the reverberation correlation interval that it must be related to the frequency of the transmitted signals by an inverse proportionality relation. This effect is in fact observed for surface reverberation, as exemplified by the oscillograms shown in Fig. 59.

The correlation interval of the reverberation envelope fluctuations may be represented approximately in the form

$$\tau_k = k_2/f_0, \tag{44.5}$$

where k_2 is a coefficient on the order of a few units, depending on the state of the ocean surface during measurement; τ_k is in seconds, f_0 in kilocycles.

We note that the correlation intervals of the reverberation envelope and phase fluctuations differ only slightly from one another, in connection with which the relation (44.5) may be assumed equally valid for the phase correlation interval.

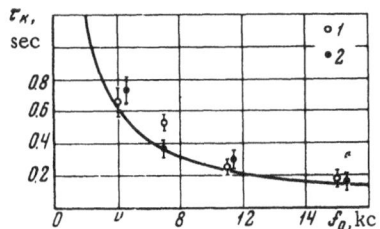

Fig. 60. Frequency dependence of the correlation interval of the reverberation envelope and phase fluctuations. 1) For the envelope; 2) for the phase. The curve represents the dependence (44.5) for $k_2 = 2.5$ kc^{-1}.

This is confirmed by the experimental data shown in Fig. 60, which also gives a curve plotted according to Eq. (44.5) for $k_2 = 2.5$ kc^{-1}; the correlation coefficient is satisfactorily approximated by a function of the form

$$R_{rE}(\tau) = r_{rE}(\tau)\cos(\Omega_r \tau), \qquad (44.6)$$

where Ω_r is determined by the wave characteristics of the ocean surface, and the envelope of the correlation coefficient $r_{rE}(\tau)$, moreover, depends on the frequency of the transmitted signals. Here the correlation interval

$$\tau_k = \int_0^\infty |R_{rE}(\tau)|\, d\tau$$

diminishes with increasing frequency according to Eq. (44.5).

For high frequencies, when

$$\tau_k \Omega_r \ll 1,$$

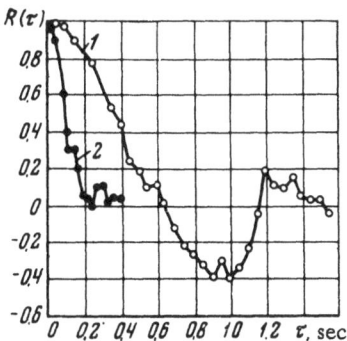

Fig. 61. Measured autocorrelation coefficients for fluctuations of the reverberation envelope. 1) For $f_0 = 4$ kc; 2) 15 kc.

the correlation coefficient (44.6) oscillates slightly, whereas, for low frequencies, the periodicity of $R_{rE}(\tau)$ is rather strongly pronounced. This is confirmed by the graph of Fig. 61.

The materials from investigations of certain statistical characteristics of reverberation in continuous transmission are contained in [30, 43], which present data on the spectral composition and envelope distributions of reverberation. In [30], in particular, it is shown that the reverberation envelope fits an almost-Rayleigh distribution, while the spectrum turns out to be somewhat wider than expected from estimates of the influence of motion of the acoustic arrays and state of the ocean surface.

A comparison of pulsed and frequency-modulated sonar systems and the investigation of reverberation during transmission of continuous frequency-modulated signals are discussed in [41, 42]. It is shown in these papers, in particular, that the reverberation created under such conditions represents a normal noise-type stochastic process with an approximately uniform spectrum, the width of which corresponds to the frequency deviation of the transmitted signals.

We point out that the theorems of superposition of stochastic perturbations make it possible in principle to calculate the statistical characteristics of reverberation for the case of frequency modulation as well, with allowance for the motion of the scatterers and acoustic arrays and with an arbitrary distribution of scatterers in the ocean. The frequency-modulation mode, like any other wide-band transmission, for example multifrequency transmission, is generally richer in information content than the tone-signal transmission mode, and in this sense the investigation of reverberation elicited by the transmission of wide-band signals should yield more refined information regarding the physical characteristics of the scatterers and the nature of sound scattering in the ocean.

§ 45. Planning and Procedure of Ocean Measurements

Experimental investigations of the statistical properties of reverberation were carried out in the period from 1955 to 1963 on the Black Sea, the Bering Sea, and the Pacific Ocean. The principal measurements were made on the Black Sea.

Two methods were used to conduct the experimental investigations, each differing from the other mainly in the disposition of the acoustic arrays, specifically as follows:

Measurements with the arrays placed at distances of from 30 to 500 m from the shoreline with signal transmission in the direction of the sea;

Measurements with the arrays mounted on structures lowered from a ship, which either stood at anchor or was allowed to drift at the time of the measurements, depending on the depth at the site.

In some cases, suspension-type rotating mechanisms, attached directly to the deck of the ship, were also used.

The experimental investigations were conducted using a special assembly of electroacoustic instrumentation, including generator devices, acoustic arrays, receiving sections, recording and monitoring equipment. For the solution of intrinsically different problems, the instrumentation complex was modified somewhat, although despite the differences in its individual components, the essential makeup of the equipment remained the same.

A typical block diagram of the instrumentation complex is shown in Fig. 62.

The transmitter included a master oscillator (1), power amplifier (2), acoustic transmitting arrays (3), and synchronizer (11).

The master oscillator formed pulsed and continuous signals with a center frequency that could be tuned within the limits from 3 to 100 kc.

In the pulsed mode signals were generated with a duration from 0.3 to 300 msec. This provided the following operating conditions:

A sinusoidal carrier and a rectangular or bell-shaped envelope;

Linear frequency modulation of the carrier with a frequency deviation up to 20 kc;

A noise carrier with the width of the frequency spectrum ranging from 500 cps to 5 kc;

The generation of a train of identical pulses for sequential transmission;

The generation of a group of pulses with different carrier frequencies for simultaneous transmission.

In the continuous mode of operation the master oscillator generated sinusoidal signals of various frequencies.

The power amplifier permitted electrical signals up to 20 kW in power in the pulsed mode and up to 5 kW in the continuous mode to be supplied to the transmitting arrays.

Transmission was realized either by plane acoustic arrays or by cylindrical arrays equipped with foam plastic reflectors, providing the required concentration and admissible side-radiation levels. The width of the directivity patterns in the vertical and horizontal planes could be varied from 5 to 20°; cylindrical-type systems, nondirectional in one of the planes, were also used.

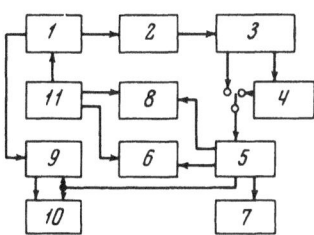

Fig. 62. Block diagram of the electro-acoustic instrumentation complex. 1) Master oscillator; 2) power amplifier; 3) acoustic arrays; 4) receiving–transfer relay; 5) receiving–amplification section; 6) monitoring system; 7) camera recorder; 8) magnetic tape system; 9) phase-meter apparatus; 10) level recorder; 11) synchronizer.

The synchronizer (11) controlled the duration of the transmitted pulses, set their repetition frequency between 0.1 and 10 cps, and executed synchronization of the receiving–transfer relay (4) when a common acoustic array (dual mode) was used in the measurements.

The receiving section included the acoustic arrays (3), amplification section (5), monitoring mechanism (6), and phase-meter apparatus (9).

The amplification section provided undistorted reception and amplification of the reverberation signals with a dynamic range of 45-50 dB.

The monitoring device included a cathode-ray indicator with amplitude deflection and audible-monitor unit with appropriate frequency converter.

The phase-meter apparatus made it possible to separate the instantaneous phase difference of the input processes and, with the incorporation of time averaging, their cross-correlation characteristics.

The reverberation signals were recorded with a photographic recording device (7), magnetic tape system (8), and level recorder (10).

The photographic device was used to record the signals on motion picture film at a film speed of up to 6 m/sec, thus providing the necessary time-wise resolving power. The magnetic tape system was capable of recording reverberation signals over the entire working frequency range.

Type N-110 level recorders and similar models were used to record relatively slowly varying processes, mainly in experiments with continuous transmission, as well as in the investigation of variations of the mean reverberation level with time.

In conducting the ocean studies, it was not necessary to perform constant absolute calibration of the electroacoustic instrumentation, since the results of processing the experimental data were usually presented in the form of normalized dependences. Only in certain cases, therefore, for example in measurements of the scattering level, was absolute calibration of the equipment carried out.

The level and form of the transmitted signals were monitored either by reception with a special hydrophone set up in front of the transmitting array or by echo signals from standard reflectors positioned at a definite distance from the arrays.

A vital factor in the performance of ocean measurements is the process of keeping check on the acoustic and hydrological conditions. In particular, in order to assess the type of reverberation being received, the nature of the sound scattering, and the characteristics of its propagation, it is necessary to perform measurements of the principal hydrological characteristics in the experimental locale.

The following characteristics were measured in almost all of the experiments under oceanic conditions:

The depth-wise temperature profiles;

The salinity of the sea water;

The form of the wave state of the ocean surface and its parameters;

The wind strength;

The velocity of the current.

In some cases, for example when the properties of bottom reverberation were to be studied, samples of the ocean bottom were also taken.

The temperature of the water was measured by means of electrothermometers at periodic intervals of several hours. The salinity, which was subject to only slight variations in any given locale, was measured once per day by taking samples of the sea water and analyzing it in the laboratory.

The characteristics of the surface wave state, as well as the sea height, average period, and swell were determined by means of conventional optical devices; the form of the waves was recorded with a specially designed electrical wave meter.

The wind velocity, velocity of the current, and type of ocean bottom were determined by means of standard meteorological and oceanographic instruments.

LITERATURE CITED

1. V. I. Bunimovich, Fluctuation Processes in Radio Receiving Devices. Sov. radio (1951).
2. L. A. Vainshtein and V. D. Zubakov, Signal Discrimination Against a Random Noise Background. Sov. radio (1960).
3. P. M. Woodward, Probability and Information Theory with Applications to Radar [Russian translation]. Sov. radio (1960). [English edition: Pergamon Press, New York (1953).]
4. S. G. Gershman and E.L. Feinberg, "Measurement of the correlation coefficient," Akust. zh., 1(4) : 326-338 (1955).
5. V. P. Glotov, "Coherent scattering of sound from clusters of discrete inhomogeneities in pulsed emission," Akust. zh., 8(3) : 281-284 (1962).
6. V. P. Glotov and Yu. P. Lysanov, "The scattered field for a spherical source above a plane layer containing discrete inhomogeneities," Akust. zh., 9(2) : 176-181 (1963).
7. G. S. Gorelik, "Theory of the scattering of radio waves by clustered inhomogeneities," Radiotekhnika i élektronika, 1(6) : 695-703 (1956).
8. G. S. Gorelik, "Influence of correlation of the scatterer velocities on the statistical properties of scattered radiation," Radiotekhnika i élektronika, 2(10) : 1227-1233 (1957).
9. I. S. Gradshtein and I. M. Ryzhik, Tables of Integrals, Sums, Series, and Products. Fizmatgiz (1962).
10. H. Cramer, Mathematical Methods of Statistics [Russian translation]. IL (1948). [English edition: Princeton University Press (1946).]
11. B. F. Kur'yanov, "Coherent and incoherent scattering of waves by a set of point scatterers distributed randomly in space," Akust. zh., 11(2) : 195-201 (1964).
12. D. Middleton, Introduction to Statistical Communication Theory, Vol. 1 [Russian translation]. Sov. radio (1961). [English edition, McGraw-Hill, New York (1960).]
13. V. V. Ol'shevskii, "Investigation of the statistical characteristics of sea reverberation," Report of the Acoustics Institute, Academy of Sciences of the USSR (1956).
14. V.V. Ol'shevskii, "Investigation of reverberation spectra in the pulsed transmission of sound," Report of the Acoustics Institute, Academy of Sciences of the USSR (1957).
15. V. V. Ol'shevskii, "Probability distribution of sea reverberation levels," Akust. zh., 9(4) : 466-472 (1963).
16. V. V. Ol'shevskii, "Correlation characteristics of sea reverberation," Akust. zh., 10(1) : 104-110 (1964).
17. V. V. Ol'shevskii, "Statistical spectra of sea reverberation," Akust. zh., 10(2) : 224-228 (1964).
18. V. P. Antonov and V. V. Ol'shevskii, "Space—time correlation of sea reverberation," Akust. zh., 11(3) : 294-299 (1965).
19. V. S. Pugachev, Theory of Stochastic Functions. Fizmatgiz (1962).
20. Yu. M. Sukharevskii, "Theory of sea reverberation due to sound scattering," Dokl. AN SSSR, 55(9) : 825-828 (1947).
21. Yu. M. Sukharevskii, "Sea reverberation in the directional transmission and reception of sound," Dokl. AN SSSR, 58(1) : 61-64 (1947).

22. Yu. M. Sukharevskii, "Sea reverberation in the presence of sound absorption," Dokl. AN
     SSSR, 58(2) : 229-232 (1947).

23. Yu. M. Sukharevskii, "Character of the fluctuations of sea reverberation," Dokl. AN
     SSSR, 58(5) : 787-790 (1947).

24. Yu. M. Sukharevskii, "Certain features of observed sea reverberation," Dokl. AN SSSR,
     60(7) : 1161-1164 (1948).

25. V. I. Myasishchev (Ed.), Physics of Sound in the Sea [Russian translation]. Sov. radio
     (1955).

26. L. A. Chernov, Wave Propagation in a Medium with Random Inhomogeneities. Izd. AN
     SSSR (1958).

27. J. O. Ackroyd, "The detection of sonar echoes in reverberation and noise," J. Brit. IRE,
     25(2) : 119-123 (1963).

28. C. G. Balachandran, "Random sound field in reverberation chambers," J. Acoust. Soc.
     Am., 31(10) : 1319-1321 (1959).

29. H. K. Carleton, "Theoretical development of volume reverberation as a first-order scat-
     tering phenomenon," J. Acoust. Soc. Am., 33(3) : 317-323 (1961).

30. P. Conley, "Continuous tone underwater reverberation," J. Acoust. Soc. Am., 27(5) : 962-
     966 (1955).

31. B. Cron and W. R. Schumacher, "Theoretical and experimental study of underwater sound
     reverberation," J. Acoust. Soc. Am., 33(7) : 881-888 (1961).

32. B. Cron and C. Sherman, "Spatial correlation for various noise models," J. Acoust. Soc.
     Am., 34(11) : 1732-1736 (1962).

33. C. Eckart, "The scattering of sound from the sea surface," J. Acoust. Soc. Am., 25(3) : 566-
     570 (1953).

34. C. F. Eyring, R. J. Christensen, and R.W. Raitt, "Reverberation in the sea," J. Acoust.
     Soc. Am., 20(4) : 462-475 (1948).

35. P. Foure, "Theoretical model of reverberation noise," J. Acoust. Soc. Am., 36(2) : 259-
     266 (1964).

36. G. R. Garrison, S. R. Murphy, and D. S. Potter, "Measurement of the backscattering of
     underwater sound from the sea surface," J. Acoust. Soc. Am., 32(1) : 104-111 (1960).

37. I. W. R. Griffiths and A. W. Pryor, "Underwater acoustic echo-ranging," Electronic and
     Radio Engineering, 35(1) : 29-32 (1958).

38. J. B. Hersey, R. H. Backus, and J. Hellwig, "Sound scattering spectra of deep scattering
     layers in the western North Atlantic Ocean," Deep Sea Research, No. 8, pp. 196-210 (1962).

39. J. B. Hersey, H. R. Johnson, and L. C. Davis, "Recent findings about the deep scattering
     layer," J. Marine Res., 11(1) : 1-9 (1952).

40. M. J. Jacobson, "Space–time correlation in spherical and circular noise fields," J. Acoust.
     Soc. Am., 34(7) : 971-978 (1962).

41. L. Kay, "A comparison between pulse and frequency-modulation echo-ranging systems,"
     J. Brit. IRE, 19(2) : 105-113 (1959).

42. L. Kay, "An experimental comparison between a pulse and a frequency-modulation echo-
     ranging system," J. Brit. IRE, 20(10) : 785-796 (1960).

43. G. A. Klotzbaugh, "Theory of continuous-tone reverberation," J. Acoust. Soc. Am.,
     25(7) : 956-961 (1955).

44. K. V. MacKenzie, "Bottom reverberation for 530 and 1030 cps sound in deep water,"
     J. Acoust. Soc. Am., 33(11) :1498-1504 (1961).

45. H. W. Marsh, "Exact solution of wave scattering by irregular surfaces," J. Acoust. Soc.
     Am., 33(3) : 330-333 (1961).

46. H. W. Marsh, "Sound reflection and scattering from the sea surface," J.Acoust.Soc.Am.,
     35(2) : 240-244 (1963).

47. J. R. Marshall and R. P. Chapman, "Reverberation from a deep scattering layer measured with explosive sound sources," J. Acoust. Soc. Am., 36(1):164-167 (1964).

48. C. M. McKinney and C. D. Anderson, "Measurements of backscattering of sound from the ocean bottom," J. Acoust. Soc. Am., 36(1):158-163 (1964).

49. L. Mittenthal, "Fluctuations of scattered noise," J. Acoust. Soc. Am., 30(9):876-877 (1958).

50. R. B. Patterson, "Backscatter of sound from a rough boundary," J. Acoust. Soc. Am., 35(12):2010-2013 (1963).

51. S. O. Rice, "Mathematical analysis of random noise," Bell System Tech. J., 23(3):282-332 (1944); 24(1):44-156 (1945).

52. R. M. Richter, "Measurements of backscattering from the sea surface," J. Acoust. Soc. Am., 36(5):864-869 (1964).

53. J. Stewart and E. A. Westerfield, "A theory of active sonar detection," Proc. IRE, 47(5):872-881 (1959).

54. W. J. Toulis, "Acoustic-wave theory interpretation of surface and bottom reverberation," J. Acoust. Soc. Am., 35(5):656-661 (1963).

55. V. M. Albers (Ed.), Underwater Acoustics. Plenum Press, New York (1961).

56. R. J. Urick, "The backscattering of sound from a harbor bottom," J. Acoust. Soc. Am., 26(2):231-235 (1954).

57. R. J. Urick and R. M. Hoover, "Backscattering of sound from the sea surface: its measurement, application to the prediction of reverberation levels," J. Acoust. Soc. Am., 28(6):1038-1042 (1956).

58. R. J. Urick and D. S. Saling, "Backscattering of explosive sound from a deep-sea band," J. Acoust. Soc. Am., 34(11):1721-1724 (1962).